Enzyme Regulation in Metabolic Pathways

Enzyme Regulation in Metabolic Pathways

Lloyd Wolfinbarger, Jr.

Professor Emeritus
Old Dominion University, Norfolk, Virginia, USA

 WILEY Blackwell

This edition first published 2017
© 2017 John Wiley & Sons, Inc.

The right of Lloyd Wolfinbarger, Jr., to be identified as the author of this work has been asserted in accordance with law.

Registered Offices
John Wiley & Sons, Inc., 111 River Street, Hoboken, NJ 07030, USA

Editorial Office
111 River Street, Hoboken, NJ 07030, USA

For details of our global editorial offices, customer services, and more information about Wiley products visit us at www.wiley.com.

Wiley also publishes its books in a variety of electronic formats and by print-on-demand. Some content that appears in standard print versions of this book may not be available in other formats.

Library of Congress Cataloging-in-Publication Data

Names: Wolfinbarger, Lloyd, Jr.
Title: Enzyme Regulation in Metabolic Pathways / Lloyd Wolfinbarger, Jr., Ph. D., professor emeritus, Old Dominion University, Norfolk, Virginia, USA.
Description: Hoboken, NJ : John Wiley & Sons, Inc., 2017. | Includes bibliographical references and index.
Identifiers: LCCN 2016053362 (print) | LCCN 2016055282 (ebook) | ISBN 9781119155386 (cloth) | ISBN 9781119155409 (pdf) | ISBN 9781119155416 (epub)
Subjects: LCSH: Enzymes–Regulation. | Metabolism.
Classification: LCC QP601.4 .W65 2017 (print) | LCC QP601.4 (ebook) | DDC 572/.7–dc23
LC record available at https://lccn.loc.gov/2016053362

Cover design: Wiley
Cover image: theasis/Gettyimages

Set in 10/12pt Warnock by SPi Global, Pondicherry, India
Printed and bound in Malaysia by Vivar Printing Sdn Bhd

10 9 8 7 6 5 4 3 2 1

Contents

Preface

This book was developed from courses in biochemistry and enzymology at the upper undergraduate and lower graduate level. I found that my students were quick at learning structures and metabolic pathways in biochemistry and at deriving rate equations for single-substrate/single-product as well as two-substrate/two-product enzyme-catalyzed reactions, but neither group of students was getting a good understanding of how enzymes directed the flow(s) of metabolites through metabolic pathways and how these metabolites regulated which pathway managed to metabolize the most metabolites to some end product(s). All too often, they were too focused on passing the tests rather than trying to understand and to challenge what I was teaching them.

Enzymology is frequently taught in a standard biochemistry textbook, and the course in biochemistry using that book takes a fairly standard approach in covering energies of activation, steady state versus quasi-equilibrium kinetics, and modes of modifier inhibition/activation of an enzymatic reaction. A typical enzymology course tends to focus on the differing kinds of enzymes and the kinds of reactions they catalyze, kinetic rate equation derivations, data plots obtained with different enzyme modifiers, and, in the more extensive texts, kinetic mechanisms by which an enzyme may catalyze a chemical reaction.

The basic premise of this book will be that there is a need for an easy means of correlating the data one actually obtains in experimental studies with multiple possible mechanisms through which some enzyme may catalyze the conversion of a substrate to a product. Although not the most appropriate means of determining some potential kinetic mechanism, quasi-equilibrium assumptions will be used throughout the book in that these keep the rate equation derivations simple. Actual metabolic pathways with known (presumed) positive and negative regulation events are linked to these potential kinetic mechanisms using both rate equation derivations and data plots illustrating how the rate equation derivations can be used to explain the data plots.

Finally, I want to emphasize that it is not the purpose of this book to be technically accurate with respect to enzyme kinetics, nor is it the purpose of this book to teach you about metabolic pathways or the traditional ways in which metabolic pathways are thought to be regulated. I do not profess to be a hard core enzymologist devoted to teaching enzymology. Rather I propose to show you how I came to understand the roles of enzymes and their kinetic constants in intermediary metabolism, and thus the purpose of this book is to challenge you to think about what I convey to you and hopefully to get you to challenge my teachings. In challenging my teachings, you will achieve the purpose of this book. You will learn to think rather than learn to take notes so you can pass the next test.

The level and contents of this book should make it suitable as a reference text in biochemistry courses at the undergraduate level and as a textbook for a course in basic enzymology taken prior to a more advanced course in enzymology at the graduate level.

I acknowledge substantial help from my former students who over my 30 odd years as a professor constantly challenged my ideas and interpretation of data, and particularly to one of my former employees, Katrina Ruth, who challenged me to stay on track with what I wanted to write and greatly assisted me in writing what I wrote in a more clear and concise manner.

Author's Review

Yes, I know that this book is repetitive, redundant, and tedious. I also know, and suggest to you, that if you expected this book to teach you about basic enzymology, you picked up the wrong book. I was a professor at a major State University for some 30 years over a time where computers were first beginning to make their appearance, and I retired when computers had pretty much taken over the thinking process once reserved for people. When I was a young Assistant Professor, I ran across a small paperback book entitled *To Know a Fly* written by Vincent G. Dethier. The book was easy to read, with cartoons no less, and described how the author learned how to ask the right questions in his conduct of research using the common housefly. He asked how did flies know when something was sweet (food), how did they land upside down on the ceiling, and other equally important questions. The answers, at least for me however, were not about learning how flies know when something is food, but rather about how to ask the right questions and come up with the proper experiments to answer a question. This is, I hope, what this book is about.

You will be dragged through pseudo-thermodynamics, complex enzyme mechanisms, rate equation derivations that go nowhere, and through seemingly endless repetition to answers that are not absolute (only possible). But, somewhere in all of this, you will hopefully begin to think about what questions to ask when it comes to the study of enzyme function(s) and how to think about the vast numbers of possibilities you can devise in the way of answers (none, of course, that will be verifiable). If you are successful, you will realize that in all of this repetition, redundancy of writing, and tedious attention to details, it is not the answers that are important in life, it is the questions that you will learn to ask.

I quickly learned to use computers and to incorporate them into my teaching. However, I never lost sight of the importance of getting a small group of people together in a classroom where the sharing of creative (and frequently "ridiculous") ideas led to understanding and knowledge with respect not only to biochemistry or enzymology, but also to all those other bits and pieces of information floating around in our heads that led us to new ideas and understanding.

I will leave you with the admonition that you should ignore the tedious, the repetition, and the redundancy and forge ahead into the book. Ignore the failings of the author and focus on creating those visual images of the "possibles" and how you arrived at those "possibles." As is always said, it is the journey that is important, not the destination.

Part I

Beyond this point there be dragons.
                    Admonition on old seafaring maps

1

Characteristics of Enzymes

At university, Enzymology was the class that most biochemistry or biology majors dreaded taking. Those students who liked the class were typically math majors who took the class for the thrill of solving complex rate equation derivations. Those students who had to take the class against their will were those who might need to understand the role of enzymes as they pertained to other aspects of biochemistry, but otherwise had little desire to sit through boring lectures involving lots of equations and the occasional molecular structures on the white board. As a professor teaching biochemistry to undergraduate and graduate students the task fell to me to keep my students' attention, so they didn't fall asleep, yet challenge them to understand why what I was teaching them could be both fun and useful.

I started teaching traditional enzymology as it was presented in the textbooks of the day (and I'm sorry to say is still being presented today). I found my students were passing the tests, but failing to understand how to interpret data and more importantly how to fit the data they were obtaining in their research into something meaningful and exciting. I eventually adopted a strategy of engaging the minds of my students with challenging, but improbable, enzymatic mechanisms and found that steady state kinetics, while more relevant, hindered the understanding of some of the more basic principles associated with enzyme kinetics. I finally hit upon the use of quasi-equilibrium assumptions and my students began to question and challenge my lectures—I had finally arrived as a professor.

Enzymology can be the study of enzymes as protein molecules with specific folding patterns of the amino acid polymer and unique binding sites wherein intra- and inter-molecular distances define the specificity of the molecule to attract, bind to, and change some substrate molecule. Computer-aided molecular modeling is a wonderful aspect of both biochemistry and enzymology in providing visuals essential to understanding, but does little to help with data analysis. Alternatively, enzymology can be the study of how these protein molecules control and mediate the flow of metabolites through intermediary metabolism affecting what we call metabolic viability to life forms. It is this latter study of enzymes that will be the focus of this book.

Enzymes are mostly proteins that are of variable length (with respect to amino acid sequences) and molecular weight. These amino acid polymers typically fold into some conformation that is most energetically favored based on the nature of

Enzyme Regulation in Metabolic Pathways, First Edition. Lloyd Wolfinbarger.
© 2017 John Wiley & Sons, Inc. Published 2017 by John Wiley & Sons, Inc.

the amino acids making up the protein and the aqueous environment in which they find themselves. For the most part, hydrophobic amino acids such as leucine or phenylalanine, as examples, are to be found in what might be called the hydrophobic core of the protein, whereas the hydrophilic amino acids such as histidine or aspartic acid, again as examples, will preferentially be found on those surfaces of the protein more exposed to an aqueous (hydrophilic) environment. The arrangement of hydrophobic and ionizable side groups of these hydrophilic amino acids is typically described as being present in some molecular organization that forms a region complementary to some low molecular weight solute. This region of the protein is generally regarded as constituting the substrate- (or modifier-) binding site. Whether this binding site tends to bind an unstable form of the substrate, stabilizing the unstable intermediate, and in so doing promoting its conversion to product; or whether this binding site tends to bind a stable form of the substrate and in so doing causes the substrate to shift into some less stable configuration promoting its conversion to product, will be discussed in detail. We will enter into this aspect of the basics of enzymology in detail in Chapter 2. For now, I only wish to stipulate that this enzyme with a substrate-binding site will be referred to as free enzyme (E) in subsequent sections. When free enzyme (E) binds with the substrate, it will be referred to as the enzyme/substrate complex (ES). As the enzyme facilitates the conversion of substrate to product via some unknown or unspecified mechanism, the product (P) released from the substrate-binding site will result in the (ES) complex reverting to free enzyme (E). Thus within the context of this book, the sum of the concentration of free enzyme (E) and enzyme/substrate complex (ES) will be referred to as the total amount (or quantity) of enzyme (E_t). When introducing modifiers of enzyme activity, I will use the simple connotation of a modifier (M) being either an activator (Ma) or an inhibitor (Mi). Modifiers will typically bind to free enzyme (E) to form a modified enzyme as either (MaE) or (MiE). Where substrate (S), enzyme (E), substrate/enzyme complex (ES), and so forth, are bracketed with square brackets, such as [S], the intent will be to express the molecule as some concentration. I will try to restate this point throughout the text, more to remind and help you than to irritate you with what will appear as my being overly redundant. Repetition is a good learning tool.

I would also like to emphasize one more point. I will make reference to "saturating" concentrations of substrate or modifier in the text. As you will see in later figures, as you add increasing concentrations of substrate (or modifier) to an enzymatic reaction, the rate of conversion of substrate to product will gradually increase until such time as that concentration approaches the capacity of that enzyme to bind to substrate converting it to product. At such a time where increasing the concentration of substrate no longer significantly increases the rate of conversion to product, it is generally assumed (described) as a saturation of enzyme by substrate. This will make more sense later, but I also want to emphasize that we will operate under the premise that the amount of substrate at any given concentration of that substrate will be inexhaustible. This means basically that you can crystallize salt out of sea water, but you will never run out of sea water where there is an infinite amount of salt. This is the difference between the concentration of salt in sea water and the amount of salt in the sea.

   I shall take a rather simplistic approach to the overall mathematical equation subject with respect to enzymes by defining a few selected terms. As you get deeper into the study of enzymes and enzymology you will have an opportunity to learn that in seeking generalities, one must frequently stretch the truth a bit in order to understand the "why" when it comes to enzymes as mediators of intermediary metabolism. I will work almost exclusively under what is generally referred to as quasi-equilibrium assumptions, rather than the more probable and ultimately more useful steady-state assumptions to describe enzyme kinetic mechanisms and associated rate constants. Later on in your studies, you can move onto steady-state assumptions, but for now I will take a bit of poetic license and work under quasi-equilibrium assumptions.

Thermodynamics

For now, let's think about the role of an enzyme and what we need to think about when it comes to an enzyme performing that role. Enzymes, as proteins in solution, have three simplistic energies. They have vibrational energy, which is simply the tendency of atoms and groups of atoms to present energy dissipation or collection as more or less a degree of stability/instability without presenting as either of the two other forms of energy. They also have rotational energy, which is simply the tendency of a molecule (collection of atoms) to "roll" or "spin" in place when in "solution." Finally, they have translational energy, which is simply the tendency of a molecule to move in some direction until events cause it to change that direction in favor of a second direction. Temperature has an impact on all three forms of energy in an enzyme, and we shall attempt to cover how all three forms of energy in an enzyme (as well as their substrates) factor into the role of an enzyme in speeding the rate of conversion of substrate to product without being consumed in the reaction. However, as I stated above, we will get more into this topic in Chapter 2. This chapter has more to do with trying to define terms than trying to explain how they help in describing how an enzyme functions.

   Temperature has an obvious role in enzyme activities and a very complex role. Temperature changes directly impact on the vibrational energy of molecules such as substrates of enzymatic reactions. Using the brief description of vibrational energy in the previous paragraph, it is easy to suggest that as the temperature in which a substrate molecule (as well as an enzyme, but let's leave the enzyme out for now) finds itself, the increased vibrational energy will tend to present as increased movement of atoms relative to their covalent bonds, as movements of electrons in possible orbits around their nuclei, and/or as overall changes in the structural conformation of the molecule (substrate in this instance). In some respects, increases in vibrational energy may represent the more significant aspect of what has been described as the "energy of activation" of some molecule necessary for that molecule to undergo a spontaneous chemical reaction becoming another molecule (perhaps a "product" for sake of my keeping in focus with this book). As a molecule becomes "activated" through the introduction of energy—in the form of increased temperature(s)—more of the substrate molecules will possess sufficient energy to acquire that "energy of activation"; and since the spontaneous chemical reaction will be defined as a concentration times some rate constant, the higher concentration of "activated" (energized) substrate will result in a faster rate of chemical change of that substrate into a product. You will encounter this issue again in Chapter 2 and Figure 2.4. However, this is a book about enzymes, and I would be remiss if I left the rather loose definition of "energy of activation" to apply only to "vibrational energy" of a molecule (substrate). Temperature also has effects on rotational and translational energies of molecules involved in some enzymatic reaction. As temperatures increase in some enzymatic reaction, molecules will tend to rotate and translate more freely, and while such rotational and translational energies may have less to do with the "energy of activation" component of a spontaneous chemical reaction, they most likely have more of an effect on the "energy of activation" associated with the enzyme-driven spontaneous chemical reaction than vibrational energy. So, how to define "energy of activation" for how we wish it to be used in the context of this book? The first thing we have to understand is that the energy of activation of a molecule that will undergo a spontaneous chemical reaction is not the same energy of activation of a molecule that will undergo a spontaneous chemical reaction where the rate of that spontaneous chemical reaction is enhanced through mediation of an enzyme (catalyst?). The energy of activation of the latter reaction should thus include roles for temperature, solution effects, enzyme, substrate, vibrational energies, rotational energies, and translational energies (note the use of "energies" here in that both the substrate and the enzyme possess these characteristics).

Like temperature, the solution in which an enzyme (and its substrates) is dissolved has an impact on all three forms of energy in an enzyme, but this impact is far more complicated than the role of temperature, although as we will see later, temperature has considerable impact on the nature of the solution and how that nature of the solution bears on the three forms of energy and subsequent interactions of the enzyme with its substrate(s) (and formed products). Enzymes function naturally in an aqueous solution of water and various ions (and/or other solutes). Water consists of a unique molecule consisting of one oxygen and two hydrogen atoms. The hydrogen atoms are situated on one side of the oxygen atom, and both kinds of atoms share electrons such that the electrons tend to spend more time with the oxygen atom than with the corresponding hydrogen atom(s) giving the whole molecule a dipole moment. This dipole moment imparts a slight negative nature to the oxygen side of the water molecule and a slight positive nature to the hydrogen side of the water molecule. It is this dipole moment that gives the water solution characteristics relevant to the energies of the enzyme, the energies of the substrate (and formed product), and directly impacts on the energy of activation and the kinetics of the enzymatic reaction in the conversion of substrate to product by that enzyme.

Normally water molecules are oriented rather randomly (with respect to their dipole moments) in some aqueous solution. However, lowering the temperature begins to remove energy from those water molecules, and at some sufficiently low temperature the water molecules will begin to lose rotational and translational energy and begin to align themselves according to their dipole moment such that the more positively charged side will be attracted to the more negatively charged side of a second molecule; eventually the water molecules will assume a "crystalline-like" structure where the dipole moments of the water molecules will all be oriented in mostly the same direction. This crystalline-like (or paracrystalline) form of water is called ice; ice presents a solution of water molecules with a lower entropy (less fluidized) than a solution of water molecules randomly associated as in a more fluidized or "liquid" solution state.

I will avoid the topic of thermodynamics, but it is important to introduce you, at this point, to a very simple equation that I think will help you to understand how the energies of an enzyme (and/or its substrate) and the solution in which it is dissolved impact the role of that enzyme. Equation 1.1 is pretty simple and describes how we currently relate entropy, enthalpy, and free energy.

$$\Delta G = \Delta H - T\Delta S \qquad\qquad (1.1)$$

The symbol Δ (delta) refers to change, G represents free energy, H represents enthalpy, T represents temperature (in degrees Kelvin), and S represents entropy. Thus, in simple terms, the change in free energy of a system equals the change in enthalpy minus the change in entropy (times the temperature). In general, if the change in enthalpy is positive the change in free energy is positive. In general, if the change in entropy is positive (no change in temperature) the change in free energy is negative. Conversely, if the change in entropy is negative (and enthalpy is constant), the system gains energy and the change in free energy is positive. If you wish to relate these events to something you are more familiar with, remember that in the universe, all the other galaxies are moving away from our galaxy as fast as they can (a large negative delta S, or an increase in entropy for the universe). I will leave you to speculate whether these other galaxies know something we don't know or whether it is this increase in entropy that facilitates the condensation of matter into a galaxy (the negative delta G for this process?). These relationships between entropy and enthalpy pertain, at least for purposes of this book, to whether or not a chemical reaction will be spontaneous or nonspontaneous, that is, whether the chemical reaction to be catalyzed by some enzyme will result in the "release" of energy (spontaneous) or require the "input" of energy (nonspontaneous) for the reaction to proceed. As a general rule, if the ΔG is positive, the chemical reaction will not be spontaneous and if the ΔG is negative, the chemical reaction will be spontaneous. The roles of ΔH (enthalpy) and ΔS (entropy) in determining whether or not some chemical reactions (and thus the enzyme catalysis of the chemical reactions) are best illustrated in Table 1.1. It is good to remember that a spontaneous chemical reaction can proceed slowly (over a very long time) or very quickly (over a very short period of time—as in a chemical explosion) and some spontaneous chemical reactions may occur so slowly as not to be measurable in the time frame under consideration. Thus in some cases a spontaneous chemical reaction may appear to be nonspontaneous and one should be careful in predicting whether or not an enzyme will actually speed the rate of some chemical reaction.

Table 1.1 The relationships between entropy and enthalpy in determining whether or not some chemical reaction will be spontaneous or nonspontaneous.

Enthalpy (ΔH)	Entropy (ΔS)	Chemical reaction is:	Relevant conditions	Comments
Positive	Negative	Nonspontaneous	Always	
Negative	Positive	Spontaneous	Always	
Negative	Negative	Nonspontaneous	Can be spontaneous	Spontaneous if temperature (T) is lowered
Positive	Positive	Nonspontaneous	Can be spontaneous	Nonspontaneous involves absolute values of ΔH and $T\Delta S$, and ΔH is greater than $T\Delta S$ Can be made spontaneous by raising temperature (T)

Now, we normally associate entropy with the degree of disorder in some system, and thus we would consider that water in a crystalline state (ice) is more ordered than water not in a crystalline state and thus this increase in order represents a solution with less entropy. It is important to understand that entropy and enthalpy in this equation are not independent when it comes to changes in free energy, but the objectives here are to try to illustrate how a water solution with more order (less entropy) will impede the "rotational" and "translational" energies of an enzyme (and its substrate) because for the enzyme to move through a solution of water molecules possessing some degree of order so as to interact with its substrate, that enzyme (and/or its substrate) will have to destabilize that ordered structure of the water solution (effectively contribute to a localized increase in entropy for the water molecules). And the only way that water can be destabilized is by contributing to a decrease in free energy, for example a negative delta G (H and T of course being held constant) or mitigation of their dipolar moments, which pretty much amounts to the same thing. This is not a difficult concept to grasp, but it is a difficult concept to present without getting deeper into thermodynamics. Suffice it to say that in examining the role of an enzyme in changing some substrate to some product it is important to understand that the nature of the water molecules in which that enzyme (and substrate) are dissolved is important. A solution of water that is more ordered will require the input of energy to make it less ordered (or more disordered) so that the enzyme and substrate can more easily approach each other and bind to permit the catalytic event converting substrate to product.

It is not just temperature that impacts on the order/disorder of the water solution in which an enzyme (and/or substrate) is dissolved. Charged atoms such as sodium (Na^+) or chloride (Cl^-) aid in the mitigation of the dipole moment of water molecules through "charge neutralization." Thus, water containing simple saline will tend to be more fluidized (more disordered) at any temperature than water not containing saline because the salt ions will tend to minimize the value of the dipole moment of a water molecule in alignment of water molecules (less disordered) according to the charge distribution on the water molecule(s). However, charged groups on an enzyme (or a substrate) may tend to organize (stabilize) the structure of the water molecules in close proximity to charged groups (especially in the absence of dipole moment mitigation by ionizable salts such as sodium chloride) on that enzyme (or substrate). Since water molecules have that dipole moment, positively charged groups on the enzyme (or substrate) will tend to order the water molecule in one direction (with respect to the partial positive/negative charge distribution of the water molecules), and negatively charged groups will tend to order the water molecules in the opposite direction (again with respect to the partial positive/negative charge distribution of the water molecules) (see Figure 1.1). This more ordered water in close proximity to a charged group on an enzyme may be thought of as "bound water" (or ordered water). A loose analogy to this bound water might include some reference to a stream of water moving between two earthen banks (as with a "river"). Water in close proximity to the bank moves more slowly than water in the middle of the stream because of the stabilizing influence (i.e., frictional resistance) of the river bank. Water along the earthen bank possesses less energy than water in the

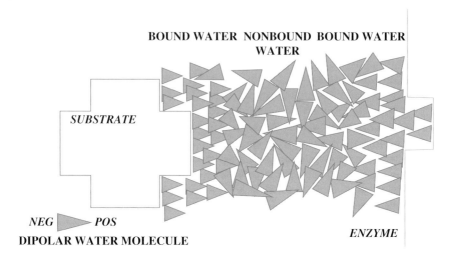

Figure 1.1 Illustration of the role of bound water versus nonbound water in the translational movement of a molecule of substrate into the substrate-binding site of an enzyme. (*See insert for color representation of the figure.*)

middle of the stream, and all things being equal is more ordered (i.e., has less entropy). The relevance of this to an enzyme-catalyzed reaction is, of course, that enzymes and substrates carry with them a layer of ordered water, and for the substrate to bind in some catalytic site on an enzyme the layers of ordered water must either participate in the binding reaction or must be disordered (increase in their entropies) for the substrate to bind in the catalytic site of the enzyme. It is easy to visualize (even if the characterization is incomplete in missing other aspects of the event) how a substrate with rotational and translational energy might give up some of that energy (less entropy for the enzyme/substrate complex) in exchange for disruption of the ordered water in some binding site facilitating binding of that substrate in the binding site. Perhaps you could imagine that the ordered (or bound) water at some enzyme binding site contributes to a reduction in the rotational (and translational) energy of the substrate as the substrate moves into the binding site and this action facilitates the binding? (Was that a good visual for the "gamers" amongst you?) If you visualized this point, you should begin to understand that binding of substrate (or modifier) in some binding site on an enzyme may be as much about the boundary water surrounding both molecules as it is about "markers" (charged or hydrophobic areas) on the substrate that the enzyme recognizes (and vice versa), and thus represent some aspect of the association "k_1" (and perhaps dissociation "k_{-1}") rate constants we will discuss in the next chapters.

We will get more into activity coefficients later in this chapter, but I would wish at this point to stipulate that with respect to activity coefficients of solutes (be they a substrate molecule or an enzyme molecule) size matters when it comes to translational and/or rotational energies. Larger molecules will present more bound water than smaller molecules, and you should therefore expect a substrate molecule to behave quite differently in solution than the enzyme molecule for a given concentration. Moreover, their concentrations (primarily increasing concentrations) will behave in a progressively less and less ideal manner—meaning their activity coefficients may change from unity (or a value of 1) at low concentrations to less than unity at higher concentrations.

This point, of course leads to an additional point (definition?) to be made in this chapter. The point is that during the course of discussions in this textbook, I will always assume that the enzyme concentration present will be at such a dilute (lower) concentration, that the free enzyme (E) will always exist as a monomer (unless I otherwise stipulate aggregation events where proteins might tend to aggregate into dimers, trimers, etc.) and will therefore behave in a more ideal state.

A third variable impacting on the order/disorder of the water molecules in which an enzyme and its substrate are dissolved is the presence (or absence) of other solutes dissolved in the water molecules (the solvent). Such solutes can, and frequently do, include ionically charged salts (sodium ions, chloride ions, etc.), noncharged molecules such as sugars, or charged molecules such as amino acids. This is, of course, why water containing such solutes "freezes" at lower temperatures than water not containing such solutes—the solutes in a water solution get mixed into the slowly organizing molecules of water as the temperature drops and negatively impact on the formation of an organized structure (crystal) of water molecules. It is also why differing ionic solutions as buffers will import differing kinetics to enzyme-catalyzed chemical reactions. I provided a bit of discussion above regarding solutes such as sodium chloride and a possible role for this salt in mitigating the dipole moment of water, but what effect(s) might nonionically charged solutes, for example a simple sugar such as glucose, have on some enzyme-mediated chemical reaction? What about a more complex sugar such as a polysaccharide (noncharged of course)? Can you look at Figure 1.1 and imagine a role for a molecule much larger than a water molecule in impacting the order of water molecules both in the bulk solution and in the bound water molecules near the boundaries of the substrate and enzyme? What if a polysaccharide were to increase the viscosity of the aqueous solution?

Enough pseudo-thermodynamics; you should be getting the idea by now that in covering the roles of kinetic constants in rate equations, the roles of pH and ionic strength, as well as the roles of groups of amino acids in forming catalytic sites of enzymes, it will be important to be able to visualize in your mind that the interaction of a substrate in a catalytic site of an enzyme involves more than that substrate simply approaching and entering (in some proper orientation) some catalytic site for it to be converted to some other chemical form (i.e., a product). What happens when a substrate molecule enters some catalytic (binding) site with the "wrong" orientation necessary for the enzyme to bind and catalyze a chemical reaction? Does a substrate have some role as a "competitive inhibitor" when it binds improperly in the binding site restricting the binding of another substrate in that binding site in the proper orientation? I'm not about giving you the answers (as one reviewer early on asked about). I'm here to try to get you to think about what answers you can come up with. It is not about your (or my) answer being "right," it is about whether or not each of us will take the time and effort to think about the question. Did we even ask the right question?

Enzyme Nomenclature

I always found discussions of enzyme nomenclature to be very dry and to be that part of the course where most students quickly lost interest in my lecture. However, enzyme nomenclature is very important when it comes to understanding the differences between enzymes, and perhaps more importantly the information is something that can be easily remembered for tests and was thus much preferred on tests than the more subjective interpretations of how some data plot could be broken down to some equation or sets of equations—the latter for which there was typically no right answer (but also no wrong answer).

The one book you will want to acquire and hold onto (if you can find one) is titled "Enzyme Nomenclature" and it contains the "Recommendations (1972) of the International Union of Pure and Applied Chemistry (IUPAC) and the International Union of Biochemistry." My worn and beat-up book was copyrighted in 1973 by Elsevier Scientific Publishing Company, Amsterdam, and has been with me for over 40 years. There was a later version published by the IUPAC in 1981: "Symbolism and terminology in chemical kinetics" in *Pure and Applied Chemistry*, volume 53, 1981, pp. 753–771, but I think more recent information is available to you. In the "modern" computer-friendly age, you will find multiple resources for enzyme nomenclature and although I do not propose to represent the utility of a given resource, nor do I recommend one resource over another resource, I would refer you to several web sites such as:

Nomenclature Committee of the International Union of Biochemistry and Molecular Biology (NC-IUBMB): Enzyme Nomenclature (www.chem.qmul.ac.uk/iubmb/enzyme/)
International Union of Biochemistry and Molecular Biology (www.iubmb.org/index.php?id=33)
Pure and Applied Chemistry: IUPAC Technical Reports and Recommendations (www.iupac.org/publications/pac/40/3/)
Science Direct (www.sciencedirect.com)
eLS citable reviews (www.els.net)

Some of these sites will require you to log on through a society membership or via other institutional login options. In some cases you are also able to purchase online access for a given period of time. The International Union of Biochemistry and Molecular Biology (IUBMB), formerly the International Union of Biochemistry or IUB, suggests that although both kinetic recommendations and enzyme nomenclature are "surprisingly" attributed to the International Union of Pure and Applied Chemistry (IUPAC) that both are in reality the exclusive responsibility of the IUBMB (but, expert chemists are, of course, consulted when appropriate).

You can, of course, get bogged down in any one of several good published narratives describing the pros and cons of how to express rate constants, concentrations of "substrates" (reactants), and concentrations of "products" in some enzymatic reaction, but for a book on the basics of enzymology, I have chosen to keep such expressions as simple as possible (it also will help you in inputting such expressions on the printed page without superscripts, subscripts, etc.—at least as much as possible). You will note, for example, that I have avoided and will continue to avoid the use of the rate constant typically called K_{cat} (or k_{kat}), which is commonly used in textbooks and rate equations. I have chosen to keep you to quasi-equilibrium assumptions and use a rate constant such as "k_2" (or similar expressions) as the rate-limiting step in the conversion of substrate to product; in defining it as the rate-limiting step it is easy for you to understand that this rate constant (or similar such expressions for a rate-limiting rate constant) define the catalytic rate (or turnover number) by which an enzyme converts substrate to product. You should appreciate, however, that when it comes time for you to publish in some journal, it will be your responsibility to read the instructions to authors and adhere to the requirements of that journal for how you describe and define any symbolism you chose to use.

Anyway, let's move on to an overview of enzyme nomenclature that you will need so as to not embarrass yourself as you participate in classes in enzymology or biochemistry and ask questions of the professor. Enzymes are typically identified by either their trivial or recommended names. The trivial names of enzymes are favored by most researchers as being more descriptive and more easily written in publications. Enzymes are typically assigned a code number (which are widely used) and such code numbers contain four elements separated by points within the following meanings:

i) The first number shows to which of the six main divisions (classes) the enzyme belongs.
ii) The second number indicates the sub-class to which the enzyme belongs.
iii) The third number gives the sub-sub-class to which the enzyme belongs.
iv) The fourth number is the serial number of the enzyme in its sub-sub-class.

I appreciate that this code system sounds awkward, but I think you will find it illuminating as we progress into the classification and coding scheme. The main divisions and sub-classes of enzymes are as follow.

1) **Oxidoreductases:** To this division belong all enzymes catalyzing oxidation-reduction reactions. The sub-class of oxidoreductases (the second number) indicates the group in the hydrogen donor that undergoes oxidation. The sub-sub-class number (the third number) indicates the type of acceptor involved.
2) **Transferases:** To this division belong all enzymes transferring a group from one compound to another compound. The sub-class of transferases (the second number) indicates the group transferred, and the sub-sub-class provides additional information on the group transferred.
3) **Hydrolases:** To this division belong those enzymes that catalyze the hydrolytic cleavage of C–O, C–N, C–C, and some other bonds including phosphoric anhydride bonds. The sub-class (second number) of hydrolases indicates the nature of the bond hydrolyzed, and the sub-sub-class (the third number) specifies the nature of the substrate.
4) **Lyases:** To this division belong those enzymes that cleave C–C, C–O, C–N, and other bonds by elimination, leaving double bonds (or conversely adding groups to double bonds). The sub-class (second number) indicates the bond

broken, and the sub-sub-class (third number) gives further information on the group eliminated.

5) **Isomerases:** To this division belong those enzymes that catalyze geometric or structural changes within one molecule (these include such generic names as racemases, epimerases, isomerases, mutases, etc.). The sub-class (second number) corresponds to the type of isomerism performed, and the sub-sub-class (third number) is based on the substrate(s) involved in the isomerization.

6) **Ligases (synthetases):** To this division belong those enzymes that catalyze the joining together of two molecules coupled with the hydrolysis of a pyrophosphate bond in ATP or a similar triphosphate. The sub-class (second number) indicates the bond formed, and the sub-sub-class (third number) is only used in the C–N ligases.

Just as an example, I will try to illustrate this numbering code for transferases by giving you a few numbering schemes. For example, transferases begin with the first number 2, and a transferase that transfers one-carbon groups would carry the second number 1. Thus you could have four sub-sub-classes within this group of transferases that transfer one-carbon groups:

2.1.1	methyltransferases
2.1.2	hydroxymethyl, formyl, and related transferases
2.1.3	carboxyl- and carbamoyltransferases
2.1.4	amidinotransferases

Finally, a transferase with the code of 2.1.1.1 is by definition a methyltransferase, and in this instance is a nicotinamide methyltransferase where *S*-adenosyl-L-methionine plus nicotinamide becomes *S*-adenosyl-L-homocysteine plus 1-methylnicotinamide.

I could go on here describing how enzyme nomenclature evolved, but I would only be reciting the words of individuals much more qualified in enzymology than I am and you would stop reading the book at this point. My purpose in giving you this information here is to help you understand some of the definitions and expressions you will encounter in the study of enzymology (and biochemistry).

Activity Coefficients

As a final attempt at "definitions" that are relevant to the study of enzymes, I would like to introduce you to the concept of an "activity coefficient" (which I alluded to earlier). I first encountered "activity coefficient" as a graduate student when I made my buffers according to the Henderson–Hasselbalch equation, but had to continually adjust the final pH of the buffer. I soon learned the difference between concentrations I measured by weight and volume and "real concentrations" I needed to achieve the desired pH. If you read Wikipedia (Wikipedia.org) you will find a definition for activity coefficient that reads, "An activity coefficient is a factor used in thermodynamics to account for deviations from ideal behavior in a mixture of chemical substances." Basically, the premise here is that in an *ideal* situation, interactions between two chemical species (an enzyme and substrate as an example) are the same (meaning, e.g., that size does not matter) and thus properties of a chemical interaction can be expressed directly in terms of simple concentrations. However, few chemical mixtures (solutions) behave in an "ideal" manner (e.g., they behave the same no matter what their concentrations, what solution(s) they are in, or what other chemicals are present in that solution) and thus deviations from ideality need to be accommodated by modifying the concentration by use of an activity coefficient. The concept of activity coefficient is closely linked to that of activity in chemistry, where activity relates to the propensity of a chemical to engage in some chemical reaction (its "effective concentration") and is derived as the product of concentration and an activity coefficient of the chemical engaging in some chemical interaction—which is, of course, what a substrate engaging with an enzyme is all about.

For purposes of this book, I will treat the activity coefficient as approaching unity. This means the value (which is dimensionless) of an activity coefficient for all reactants being discussed will be equal to 1, and therefore any concentrations discussed will represent the effective concentration of that reactant. This is not the time to get into such details regarding how the concentration of a substrate may not be the effective concentration of that substrate; however, you will need to be aware of the differences between a concentration and an effective concentration when it comes to dealing with enzyme kinetics and rate constants in your research. If you wish to obtain specific activity coefficients I would refer you to http://sites.google.com/.../activity-coefficient/. However, you will be wise to understand that activity coefficients of a chemical (enzymatic) reactant can change relative to the concentrations of the reactants in the chemical and/or enzymatic event, and reactants at very dilute concentrations may have activity coefficients approaching unity, but very different activity coefficients at high concentrations. This is why enzymology (working with enzyme kinetics and reaction rates) is difficult.

2

Self-Assembly of Polymers

Although small children have taboos against stepping on ants because such actions are said to bring on rain, there has never seemed to be a taboo against pulling off the legs or wings of flies. Most children eventually outgrow this behavior. Those who do not either come to a bad end or become biologists.

Vincent G. Dethier, *To Know a Fly* (Holden-Day)

This is a difficult chapter to cover a topic not often covered in textbooks on enzymology and I will thus endeavor to keep it short. I included this chapter because it is important for you to understand that it is a form of self-assembly that leads a substrate molecule to bind into the substrate-binding site on an enzyme. Self-assembly of polymers represents those sorts of molecular interactions that lead to a polymer where the units of that polymer contribute to the assembly of that polymer and make it possible for an enzyme to form covalent bonds that yield the "final" polymer. Before we get into polypeptides/proteins, let me first ask if you know why nucleic acids occur as double helix structures? It was always an easy thing to teach my students that DNA (and parts of RNA) were double helical structures, but when I would ask why they were double helical structures very few students could tell me the simple reason for why they are. Indeed, more than once I was told that single-stranded nucleic acids were not helical in structure. Let me also ask if you ever wondered how/why complex organic molecules formed in some "primordial soup" that eventually and inevitably led to polymers such as double-helical deoxyribonucleic acids, carbohydrates, and proteins? It wasn't because they arrived on this planet on a meteor. Neither did they form via some enzymatic mechanism. So what was the driving force(s) that led to the appearance of complex organics and polymers possessing "enzymatic" activities? I'll give you a hint: It is the same driving force(s) that lead to substrates combining with an enzyme and subsequently being changed to a product.

We tend to accept that in the early formative years of this planet, volcanic outgassing led to large amounts of water vapor, hydrogen sulfides, carbon dioxide, various forms of nitrogen (nitric oxides, cyanides, ammonia, etc.) and any number of noxious and foul-smelling gases. Indeed, we still get such outgassings from volcanoes today. Furthermore, we are typically informed that as the earth cooled, the water vapors condensed and the volatile gases dissolved in that water.

Enzyme Regulation in Metabolic Pathways, First Edition. Lloyd Wolfinbarger.
© 2017 John Wiley & Sons, Inc. Published 2017 by John Wiley & Sons, Inc.

If you can accept, or have already accepted, these points, then you are ready to move on to how the formation of "life" on Earth was not only probable, but inevitable.

Given the complex solutions of dissolved gases in water, all we needed was some mechanism to drive chemical reactions wherein the dissolved gases condensed into more complex structures. What we needed was some means to make delta *G* (where G is usually defined as "Gibbs free energy" and "delta" implies change) negative, and we can have that if some overall driving change in entropy occurs, and we can have that if the water evaporates (dramatic increase in the entropy of the water). This evaporation not only concentrates the dissolved gases (mostly simple inorganics), but provides a driving force (among others) to otherwise reluctant condensation reactions. Indeed, if you look at all important polymerization reactions (important to the formation of "life"), you will note that sugars are basically repeating units of carbon dioxide/carbon monoxide (in a more reduced state) where a molecule of oxygen (in a more reduced state, i.e., water) is lost. So, in simple terms, sugars are formed from carbon oxides dissolved in water through the loss of water from the reaction—a "dehydration" event (under reducing conditions). Indeed, the formation of a polysaccharide from sugars is also a dehydration event where the loss of a molecule of water results in the condensation of two sugars to form a disaccharide and eventually polysaccharides (branched and not branched). It sort of comes down to how much water needed to "evaporate" from the solution of dissolved gases to achieve ("drive"?) the desired (?) chemical reactions (Figure 2.1).

If we move onward, we can consider more complex molecules such as adenine (of nucleic acid fame). Adenine is basically multiple molecules of cyanide (a carbon and nitrogen atom grouping). Cyanide is a molecule consisting of carbon and nitrogen bound by a triple bond and carries a net negative charge; it is thus typically present in association with a hydronium ion (positively charged hydrogen) to give hydrogen cyanide in water solutions. Typical bases found in nucleic acids are comprised of structures easily represented by five molecules of hydrogen cyanide condensed into a complex molecule presenting electronegative clouds of electrons above and below the ring structure (Figure 2.2). It is these

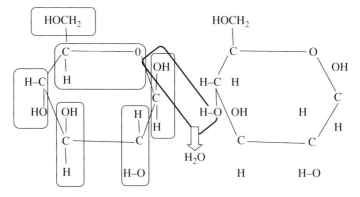

Figure 2.1 Sugar comprised of five molecules of "CO_2" covalently linked to a second sugar with loss of a molecule of water.

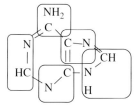

Figure 2.2 Illustration of the use of five molecules of cyanide to form the base adenine.

electronegative clouds of electrons that contribute to the stacking tendency of nucleic acid bases in an aqueous environment.

Without going into this further, let's try to move on to why nucleic acids are helical as polymers and why (how) such polymers might have formed. The point I wish to make so far is that simple gases could easily have condensed (via simple dehydration events) to form the more complex organics needed to make those molecules necessary for life as we know it today.

Nucleic acids are comprised of phosphate (H_2PO_4), the sugar ribose, and an assortment of purines (adenine and guanine, specifically) and pyrimidines (cytosine and thymine/uracil, specifically) (Point: these exact bases may or may not have been present in any original polymeric "nucleic acids," but this is not a book on molecular evolution). When you dissolve purines and/or pyrimidines in water, they tend to stack like stacks of pennies. This stacking propensity (or property) has to do with electron clouds above and below each molecule, and the more concentrated these molecules (pennies) become, the longer (taller if you prefer) these stacks become. Thus, it is a natural tendency of the bases found in nucleic acids to stack in a linear fashion. They self-assemble themselves into the basic structure that is an essential component of their function in life. One linear polymer will also serve to align matching bases, adenine to thymine and guanine to cytosine, creating a matching polymer oriented in the opposite direction of the first polymer. If we introduce phosphates and sugars and encourage a series of "dehydration" events we can basically remove water molecules between a phosphate and a sugar and water molecules between a sugar and a base (a purine or pyrimidine); hence we are able to see that formation of these covalent linkages locks the double-stranded helix into the forms we know as deoxyribonucleic acid or ribonucleic acid. That there are 10 base pairs per turn of the helix is dictated by the bond lengths and the attractive and repulsive forces inherent in the molecules comprising the polymer. Thus, polymerizations of purines and pyrimidines into nucleic acids are quite analogous to the polymerizations of sugars into polysaccharides. When we factor in the tendency of the bases to stack tightly and the phosphates (with their strong electronegative charges) to repel each other, we observe that we have two opposing forces within this polymer. The base stacking tends to compress the polymer and the phosphates tend to elongate the polymer. The easiest way to satisfy these two opposing forces is to make the point that if the base stacking can be shifted to either the left or right along the linear run of the polymer, one can get greater physical separation of the phosphates from each other yet allow the bases to still stack tightly. Nucleic acids are helical due to

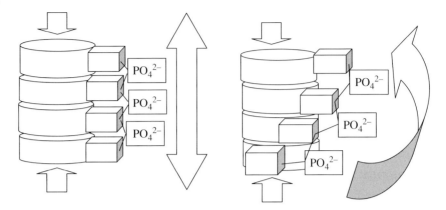

Stacked Bases/Ribose/Phosphate   *Rotated Stacked Bases/Ribose/Phosphate*

Figure 2.3 Cartoon to illustrate the roles of base stacking and strong electronegative repulsion in determining why nucleic acids are helical.

these opposing forces, and the helical nature is energetically the most favorable structure the polymer can assume. The sums of the parts dictate the structure of the whole (Figure 2.3).

I suspect by now you are wondering what all of the above has to do with enzymes. I have taken considerable liberties with you in order that I might make the point that chemistry was the basis for the formation of more complex molecules and that if you shake a concentrated solution of phospholipids, proteins, carbohydrates (polysaccharides), nucleic acids, and so forth you will gain membrane-bound structures that will stain nicely using the simple Gram stain used to distinguish Gram-negative versus Gram-positive microorganisms and look like a typical bacterium under the microscope. The basic ingredients dictate the complex structures they can become without the role of enzymes.

So, we get to the chemical formation of proteins. The point to be made essentially echoes the points I made earlier, namely that amino acids are comprised essentially of "carbon dioxide" and nitrogen (in various associations with hydrogens and/or oxygen). Indeed, the simplest amino acid, glycine, is comprised of nitrogen, a reduced carbon, and an oxidized carbon—namely "carbon dioxide"—and proteins are simple polymers of amino acids condensed together via some form of a "dehydration" event (Figure 2.4). It is possible to create polypeptides with differing characteristics (and even crude enzymatic activities) by controlling the numbers and kinds of amino acids you condense (again a basic "dehydration" reaction) into a polymer.

Polymeric amino acids comprised of large number of hydrophilic basic amino acids (such as arginine or lysine) present uniquely differing properties than polymeric amino acids comprised of large numbers of the hydrophilic acidic amino acids (such as aspartic acid or glutamic acid). Both are usually very soluble in water solutions whereas polymeric amino acids comprised of large numbers of hydrophobic amino acids (such as phenylalanine, leucine, isoleucine, tryptophan, tyrosine, etc.) are usually not very soluble in water

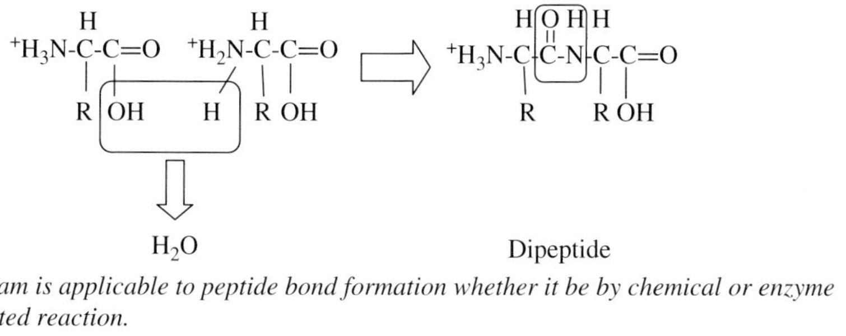

Diagram is applicable to peptide bond formation whether it be by chemical or enzyme mediated reaction.

Figure 2.4 Illustration of the role dehydration has in the formation of peptide bonds in proteins. The diagram is applicable to peptide bond formation whether it be by chemical or enzyme-mediated reaction.

solutions. It is pretty easy to visualize how clay particles (or any other kind of particle) might have assisted in the attraction of amino acids with mostly electronegative, electropositive, or hydrophobic characteristics to their side groups onto their surfaces such that as those particles dried out, selective dehydration/condensation reactions would have led to polymers of amino acids with unique properties. During a rehydration event, such polymers would have been expected to "fragment" to some extent, but participate in another dehydration/condensation event eventually leading to polypeptides possessing a wide variety of distributions of amino acids and associated physical characteristics. When a polypeptide contains all such kinds of amino acids, the nature of the amino acids, the relative amounts of hydrophobic and hydrophilic amino acids, the distributions of these amino acids as well as the pH and ionic charge distributions on the ionizable groups associated with various amino acids will determine the folding patterns of the polypeptide (perhaps we can call it a protein now?). And, of course, the folding patterns of the protein will determine secondary and tertiary structures (and the packing density of the polymer) as well as which amino acids are exposed in some region we might think of as being a substrate (or a modifier) binding site. We should also assume that the binding of some substrate (or modifier) to the protein might or might not cause changes in the folding patterns of that protein. Again, the sums of the parts dictate the function(s) of the whole.

So why might we be interested in the folding patterns of proteins and/or changes in that folding pattern due to binding of some small molecular weight metabolite somewhere in association with the protein? In very simplistic terms it comes down to whether or not an enzyme speeds the rate of some chemical reaction either by binding to and stabilizing some unstable intermediate form (an "activated" form) of a substrate (effectively increasing the concentration of this unstable intermediate and thus increasing the rate of the otherwise chemical reaction), or by binding to a stable form of a substrate and in so doing undergoing a structural change that makes the stable form of the substrate unstable (pardon the run on sentence). These two simplistic mechanisms by which an enzyme may speed the rate of a chemical reaction are obviously not the only possible mechanisms, but then this book is less about trying to be all inclusive than about getting you to visualize ways in which you think enzymatic mechanisms may serve to speed chemical reactions. If we keep this book simplistic, you will fill in other options after you find reasons to disagree with my simplistic ideas and perhaps forgive me for taking liberties with the details.

Let's try to think in a bit more detail about the two simplistic mechanisms I have proposed to you above. First of all let's think about some population of substrate molecules in an aqueous solution. Within the total population of substrate molecules, each molecule will possess a different level of energy (be it vibrational, rotational, or translational). One can easily accept that the distribution of substrate molecules with differing levels of energy will be rather uniformly distributed around some "average" energy level and that the vast majority of substrate molecules in the total population will possess some "average" energy level. This distribution will leave small numbers of substrate molecules with lower energy levels and small numbers of substrate molecules with higher energy levels. We generally assume (expect) that substrate molecules with some higher level of energy will be more unstable than the rest of the substrate molecules, meaning the energetics of one or more bonds in these molecules will be more likely to undergo some physical change. These are the molecules generally assumed in textbooks to be "activated" (i.e., achieve the "energy of activation" levels necessary for the chemical reaction to proceed spontaneously). We can move the overall energy distribution profile of the total population of substrate molecules by either removing energy or adding energy to the system (cooling or heating the solution, respectively). Obviously, at some temperature, we will move the "average" energy level to a higher state of energy thus dramatically increasing the concentration of the unstable form(s) of that substrate, and the chemical reaction will proceed more quickly. This is the point in enzymology where we are told (I was told) that enzymes "lower the energy of activation" in a chemical reaction, but we never seem to get the opportunity to talk about how an enzyme might lower the energy of activation of an otherwise spontaneous chemical reaction (which I never understood). In this book I have attempted to provide two options to try to help you understand how an enzyme may help to speed the rate of a chemical reaction. Option one has the enzyme binding to the unstable intermediate thus preventing/hindering the unstable intermediate from its equilibrium shift back to the stable form (i.e., not activated) of the substrate and thus effectively increasing the overall concentration of the unstable intermediate. You will note that this option does not involve a lowering of the energy necessary to achieve a state of being "activated"—only some change in concentration (an increase) of activated molecules. Option two would be where the enzyme binds the stable form of the substrate and undergoes a conformational shift in structure due to masking of groups on the enzyme; this shift in structure induces a change in the molecular structure of the substrate, changing it to the activated form and thereby achieving the same kind of increase in overall concentration of the unstable form of the substrate. We could also consider at this point that in binding to the enzyme, ionic interactions of the enzyme with the substrate could lead to a destabilization of bond lengths/strengths, which could also be considered as converting an otherwise stable molecule (substrate) to an unstable (or activated) molecule. Both options will, of course, lead to an effective increase in the concentration of the unstable form of the substrate and thus an increase in the rate of chemical conversion of that substrate to a product. Again, there is no lowering of some "energy of activation" in this (these) option(s) either. There are, of course, sub-options to Option Two, where, for example, simply binding of a substrate to a site on the enzyme aids in the destabilization of the substrate without conformational changes to the enzyme, perhaps by providing for a hydrophobic domain that could aid in some "dehydration" event, but you probably already thought about that if you have read anything about how ribosomes aid in the formation of peptide bonds. Does this "sub-option" lower the energy necessary for the molecule to be "activated"?

We thus come to an interesting question, which I will not try to answer, but if we consider the two possible mechanisms from above (that the enzyme binds either the stable or unstable form of the substrate), is there an opportunity for you to determine whether either (or none) of the two mechanisms are at play by examining the effect(s) of temperature on the rate(s) of an enzymatic reaction? If the enzyme binds to the stable form of the substrate molecule, small changes in the population of unstable (or active) molecules may not dramatically change the concentration of the substrate molecule that binds to the enzyme and increased temperature might even reduce the concentration of the stable form of the substrate that actually binds to the enzyme. However, if the enzyme binds to an unstable form of the substrate, raising the average energy in the population of molecules would be expected to increase the concentration of the unstable forms of the substrate in the total population of the substrate (Figure 2.5). Perhaps even

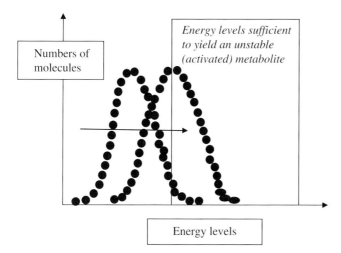

Figure 2.5 An "X/Y" data plot to illustrate two differing groups of molecules possessing two differing average energy levels permitting the demonstration of how increasing the average energy levels of such a population will impact on the concentration of substrate molecules possessing sufficient energy to be "unstable" and thus chemically reactive. The graph is designed to illustrate the population distributions of substrate molecules with differing levels of energy where the average energy levels are present at two different values. The vertical line illustrates the numbers of molecules presenting sufficient energy to be described as being "unstable" and likely to chemically change from one molecular form to another molecular form, that is, undergo conversion from a "substrate" to a "product" as described later in this book.

to the point where all the substrate molecules are in the activated or unstable form and thus at very high concentrations. Is the experimental design too complicated by effects of temperature on the enzyme as well as the substrate to make this experimental design possible? After all, we should expect that increasing temperatures will change the folding patterns (and thus any binding and/or catalytic site) of the protein as well as have direct effects on some spontaneous chemical reaction converting substrate to product without action by the enzyme. It is good to think about such things.

We should not, of course, ignore the option(s) of how the substrate bound in the substrate-binding site (perhaps the catalytic site?) might be more or less likely to be released (the "[ES]k_{-1}" part of the equilibrium binding of substrate to enzyme illustrated in Chapter 3 and elsewhere in this book). If an enzyme binds the stable form of substrate and activates the substrate to a less stable form through some conformational or ionic shift in bond angle/strength, is the "substrate" being released in the [ES]k_{-1} kinetic step the stable or unstable form of the substrate S? You will note that the substrate S in the equations in Chapter 3 does not make reference to whether the substrate is "S" or "S*" (where S* might be used to indicate the activated or unstable form of the substrate). Does it make a difference? Could you use a similar experimental design as described above to determine the effect of changing the overall temperature(s) at which your kinetic studies were performed to determine whether or not S versus S* was the substrate in question in the equations?

Let's think about what we have talked about so far. We have considered that the fundamental interaction of some substrate (or modifier) molecule must approach some binding site on an enzyme. That in approaching the enzyme, the substrate must disorder bound water molecules on both itself and the binding site on the enzyme. That this disorder can be achieved using both (mostly) the rotational and translational energies of the substrate. However, we must also think about what is the energy form of the substrate that will most effectively (efficiently) bind in the binding site of the enzyme. Is it the stable or unstable form (species) of the substrate? If it is the unstable form of the enzyme, temperature may have a decided impact on the binding simply by making the effective concentration of the unstable form (species) greater (not to mention destabilizing the ordered water molecules in the boundary zones of both the substrate and enzyme), thus impacting directly on some rate constant, to be later defined as k_1, since the effective substrate concentration (be it the stable or unstable form of the molecule) has a direct effect on a rate (or velocity of the reaction event) as defined by:

$$\text{Velocity/rate} = [S]k_1$$

However, we should not forget that raising the temperature of the system (increased energy) will also tend to destabilize the water molecules making up the bound (or if you wish, thinking of it as "barrier" water is also quite appropriate), further enhancing negativity to Gibbs free energy (i.e., a more negative delta *G*). If all this sounds confusing, remember it is important for you to remember that many things participate in a chemical and enzymatic reaction leading to a change in some molecule to another molecule, and it will be your role to try to find as many such things as possible.

You will be introduced to some simple rate equations in Chapter 3 and you will eventually want to try to consider any and all such things that can come into play as you begin to think of an enzymatic reaction as defined by rate constants and concentrations of reactants. You will particularly want to think about these "things" as you get into how modifiers may modify enzymatic mechanisms, as described in Chapter 5.

3

Beginnings of Equations

All we have to decide is what to do with the time that is given to us.
Gandalf the Grey in J.R.R. Tolkien's *The Lord of the Rings*

It is with apologies that I get so quickly into an equation; however, please be assured that I do so only to set the stage to understanding why I make the point about stretching the truth. I am about to define two time-honored values, K_m (or Michaelis constant as it is frequently called) and V_{max} (which is generally thought of as the maximal rate at which a given amount of an enzyme E may convert substrate S to product P). In practical terms, the Michaelis constant (or K_m) is equivalent (note the use of "equivalent" and not equal) to that substrate

Illustration 3.1

Enzyme Regulation in Metabolic Pathways, First Edition. Lloyd Wolfinbarger.
© 2017 John Wiley & Sons, Inc. Published 2017 by John Wiley & Sons, Inc.

concentration at which a given quantity of enzyme can convert substrate to product at half of the maximal rate possible. For purposes of this book, Equation 3.1 below is going to define K_m as a ratio of those rate constants (k_1, k_{-1}, and k_2) where those rate constants leading to the dissociation of the enzyme/substrate complex (ES) are divided by the rate constant leading to the association of the enzyme/substrate complex. You will find continual reference to this particular equation in later chapters, where we discuss how modifiers are used in management of intermediary metabolism where such modifiers alter the apparent K_m and/or apparent V_{max} values of an enzyme. We will also discuss what "apparent" means in this previous sentence.

$$K_m = \frac{(k_2 + k_{-1})}{k_1} \tag{3.1}$$

Suffice it to say that this is just one of many "definitions" of K_m and the best overall definition of K_m for an enzymatic reaction is: that substrate concentration which will provide for a measured velocity of conversion of a "substrate" to a "product" which is half of the maximal rate possible. You will see Equation 3.1 a lot in studying enzymology. It forms the basis for the Michaelis–Menten mechanism and assumes that the enzyme–substrate complex is in thermodynamic equilibrium with free enzyme and substrate, but is only true if k_2 is much much less than k_{-1} (and k_1 for purposes of this book). This relationship of rate constants is usually expressed as: $k_2 << k_{-1}$ (k_1), and the use of "much much less than" in text form will be used interchangeably with "<<" in this book (the "much much" is not a typographical error).

The thermodynamic equilibrium assumptions that form much of the discussion and equations in this book come as an extension of the original Michaelis–Menten mechanism where the K_s in their mechanism will correspond to k_1 and K_{cat} will correspond to k_2 (and where I will use a dissociation constant k_{-1} rate constant). In general we will define an enzymatic reaction by the simple equation:

$$S + E \underset{k_{-1}}{\overset{k_1}{\rightleftharpoons}} ES \xrightarrow{k_2} P + E \tag{3.2}$$

where k_1 is $\geq k_{-1}$ and both k_1 and k_{-1} are much much (<<, as above) greater than k_2 (note the "much much greater than" terminology).

Using the information in Equation 3.2 and the conditions utilized to define the relationships of the individual rate constants (i.e., k_1, k_{-1}, and k_2) we may now define V_{max} (as we propose to use it in the context of this book). The conditions set in Equation 3.2 above enable me to make a couple of points. Firstly, that by setting k_2 as being much much less than k_1 and k_{-1}, I have made the conversion of the ES complex to product (P) and free enzyme (E) as the rate-limiting step (K_{cat} in other textbooks) in the overall conversion of substrate to product. Secondly, I have also made k_1 greater than or equal to k_{-1} and thus I have stipulated that the enzyme–substrate complex (ES) is almost as likely to be formed from substrate (S) and free enzyme (E) as the enzyme substrate complex is to break down into substrate (S) and free enzyme (E). You have probably understood by now, that in the grand scheme of converting substrate to product, the interconversions between substrate and enzyme and the enzyme–substrate complex will come to something approaching equilibrium before a significant amount of the enzyme–substrate complex (ES) breaks down to product (P) and free enzyme (E). This situation is what has been called quasi-equilibrium assumptions for determining the mechanism(s) for converting substrate to product. You will learn more about other such mechanisms, that is, steady-state kinetics, and so forth—where k_2 may not be much much less than k_{-1} (and/or k_1) and where some aspect of the reaction is deemed to be at a state where that aspect is appearing at approximately the same rate that it is disappearing—when you get further along in this book or are unfortunate (fortunate?) enough to have to take a real class in enzymology one day. However, for now we shall operate under quasi-equilibrium assumptions for purposes of defining K_m and V_{max} values, in that the purpose of this book is to help you to understand better how enzymes manage/control the flow of metabolites through metabolic pathways and how these metabolic pathways are manipulated within a cell to control which pathways metabolites are directed into and into which products these metabolites are eventually converted.

As you will observe later, I will tend to talk about velocity (v) and/or maximal velocity (V_{max}) as being analogous to the inside diameter of a piece of pipe that can be used to carry water, that is, where the inside diameter of that pipe represents the water-carrying capacity of that pipe in a manner similar to the substrate-carrying capacity of a given quantity of enzyme (the more enzyme the bigger the pipe) in converting substrate to product. Such analogies are not always accurate, but it is hoped that they will provide visual images that will permit you to better understand how the K_m and V_{max} attributes of an enzyme can be used to control the movement of metabolites through metabolic pathways and how by modification of these same K_m and V_{max} values the cell can manipulate where it directs these flows of metabolites and thus the products that are made via these metabolic pathways.

Under the terms and conditions of Equation 3.2, we may thus write an equation defining velocity (or rate) of conversion of substrate (S) to product (P):

$$\text{Velocity}(v) = k_2([ES]) \tag{3.3}$$

where velocity (v) describes the measured rate at which substrate (S) is converted to product (P) as being equal to the rate constant k_2 times the concentration of the enzyme–substrate complex ([ES]).

It should be obvious to you that when the concentration of the enzyme–substrate complex ([ES]) approaches being nearly equal to the total concentration of enzyme ([E_t]), that the measured velocity (rate) of the conversion of substrate (S) to product (P) will occur at approximately the maximal rate possible:

$$\text{Maximal velocity }(V_{max}) = k_2([E_t]) \tag{3.4}$$

where maximal velocity (V_{max}) describes the maximum possible measured velocity (rate) of the conversion of substrate (S) to product (P), k_2 is the rate constant for the conversion of the enzyme–substrate complex (ES) to product (P) and free enzyme (E), and [E_t] defines the effective concentration of total enzyme present in the system.

It should also be obvious to you that [ES] can never equal [E_t] (where [E] + [ES] = E_t) and thus one will always be left with the situation where maximal velocity may be approached, but never achieved. This observation is best illustrated (Figure 3.1) utilizing the time-honored velocity versus substrate concentration curve, traditionally shown in textbooks and drawn on white boards during lectures on enzyme kinetics.

We may also take this opportunity to generate an equation you will need over and over as you study enzymology. Equations 3.1, 3.3, and 3.4 form the basis of the Michaelis–Menten equation:

$$v = \frac{[E_t]+[S]k_2}{K_m +[S]} \text{ OR } v = \frac{V_{max}[S]}{K_m +[S]} \tag{3.5}$$

Equation 3.5 can be illustrated graphically in Figure 3.1, where both K_m and V_{max} can be visualized as a function of velocity of the enzymatic reaction versus substrate concentration. It should be emphasized here, that where substrate concentration is considered, you should understand that whatever the concentration of substrate, there is an infinite amount of substrate such that the concentration never gets reduced by the conversion of that substrate to product.

What does the term "asymptotically," as used in the legend to Figure 3.1, mean? In simple terms, it means that as you leave to go home tonight, in the first 15 minutes you get half-way home. In the second 15 minutes you get half-way home again. In the third 15 minutes you get half-way home again. This goes on and on until your nose is almost touching the front door. You asymptotically approach

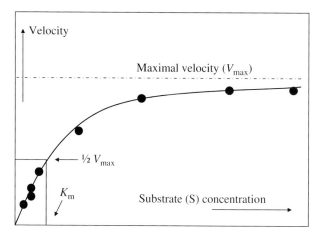

Figure 3.1 Typical velocity versus substrate concentration data plot showing the measured velocity asymptotically approaching maximal velocity at infinitely high concentrations of substrate. (*See insert for color representation of the figure.*)

getting home, but you never quite get there. Eventually you are so close to home that for all practical purposes you are home, but "just not there." The data plot illustrated in Figure 3.1 simply states that the measured velocity of an enzymatic reaction at infinitely high concentrations of substrate will result in almost all free enzyme being converted to an enzyme–substrate complex, but "just not there."

Figure 3.1 also provides an illustration of what K_m (the Michaelis constant) corresponds to on a typical velocity versus substrate concentration data plot. In that I have already defined the K_m value as being equivalent to that substrate concentration that provides for a measured velocity equal to one-half of the maximal velocity possible, one can simply draw a horizontal line in Figure 3.1 where the measured velocity of the enzymatic reaction is half of the V_{max} such that it intersects with the line forming the data plot and then dropping a line vertically down to the substrate concentration needed to achieve that specific measured velocity. In that the measured velocity is asymptotically approaching V_{max}, one does not know exactly the velocity of V_{max} and thus one does not know exactly where one-half V_{max} occurs on the data plot. Thus, it is difficult to find that velocity value where a substrate concentration would be sufficient to achieve one-half of the V_{max}.

Figure 3.2 illustrates a second way to more accurately calculate both K_m and V_{max} than in Figure 3.1, where one has conducted a series of experiments at a given concentration of enzyme to determine the change in concentration of substrate (S) or product (P) over time at several different substrate concentrations. It is important to emphasize here that these concentration changes versus time at the various substrate concentrations must be linear such that one is always measuring conversion of substrate to product as a first-order reaction (see Figure 8.2, later). I have chosen what is typically referred to as a Lineweaver–Burk (or double reciprocal) data plot to illustrate this better way to calculate K_m and V_{max} values, not because it is the best or more accurate than other methods,

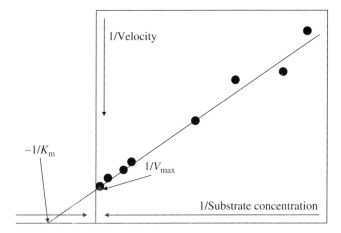

Figure 3.2 Typical Lineweaver–Burk data plot where the reciprocal of the velocity at the reciprocal of each substrate concentration is plotted. The values of V_{max} and K_m are calculated by extrapolation of where the data line crosses the *x* and *y* axes for 1/Velocity and 1/Substrate concentration.

but rather because everyone is more familiar with it than the other means of plotting data (e.g., the Eadie–Hofstee plots). As a data plot, the Lineweaver–Burk plot places too much emphasis on measured velocities at the lower substrate concentrations where such measured velocities are expected to present the greatest error in the determination of changes in substrate or product concentration over time. This overemphasis tends to skew the linear plot more than similar data points (velocity values) obtained at the higher substrate concentrations, and thus the Lineweaver–Burk means of plotting these kinds of data tends to contribute significant error to the calculation of both K_m and V_{max} values, with greater error potential directed toward calculation of K_m values. However, as already indicated, although data plots such as the Eadie–Hofstee are more accurate than the Lineweaver–Burk plot, I feel that the ease of use and familiarity of most people with data plotted using a Lineweaver–Burk plot format outweigh such disadvantages. I would also add that it is not really the objective of this book to get into accurate determinations of such values for enzymes, but rather to sufficiently familiarize the reader with what these terms mean relative to the role of enzymes in metabolic pathway regulation so that the reader is able to grasp the information provided in later chapters.

Here I only need to make one additional point with respect to K_m values. I chose to wait until this point because I did not want you to confuse a K_{eq} (or "affinity constant") with K_m, as is so frequently done. If you go back to Equation 3.1 you will note that the rate constant k_2 is present in the denominator and thus, although very small relative to both k_1 and k_{-1}, contributes to the value of K_m. Under those circumstances where the enzyme–substrate complex (ES) does not break down to product (P) and free enzyme (E), there would be no k_2 value and Equation 3.1 would become Equation 3.6:

$$\text{Kequilibrium}\left(K_{eq}\right) = \frac{k_{-1}}{k_1} \tag{3.6}$$

K_m would now be K_{eq}, and K_{eq} would represent a ratio of the association/dissociation constants and a measure of the "relative affinity" of the enzyme for the substrate molecule (S) in the binding site of the enzyme (E). Because k_2 values in this mechanism have already been described as being much much less than either k_1 or k_{-1}, the K_m value of the enzyme (or any enzymes) described in subsequent chapters would be very similar to a K_{eq} for those enzymes. Thus for the purposes of this book (and for this book only) I will use K_m as a relative measure of the affinity of an enzyme for its substrate.

I will also take poetic license and stipulate that we will focus on K_m values ranging over a couple of orders of magnitude, for example negative 6 logs and negative 8 logs. For ease of understanding, the reader may think of a negative 6 log K_m value as requiring a substrate concentration of 1×10^{-6} M to achieve a measured velocity in the conversion of substrate to product equal to ½ of the V_{max}. Conversely, the reader may think of a negative 8 log K_m value as requiring a substrate concentration of 1×10^{-8} M to achieve a measured velocity in the conversion of substrate to product equal to ½ the V_{max}. It should be obvious that an enzyme with a negative 8 log K_m value will require at least a 100-fold lower concentration of substrate to achieve ½ V_{max} than an enzyme with a negative 6 log K_m value. I have emphasized this here in order to point out, in no uncertain terms, that an enzyme with a K_m of 10^{-8} M is an enzyme with a relatively greater "affinity" for the substrate than an enzyme with a K_m of 10^{-6} M. Also, remember that with Lineweaver–Burk data plots, the data being plotted are reciprocal values, and larger values of velocity or substrate concentration are those data points approaching the origins of the x and y axes of this data plot. Finally, remember, that under the current assumptions, k_1 and k_{-1} contribute only minimally to the overall conversion of substrate to product, in that k_2 is very small compared to them and is thus the rate-limiting step in the overall velocity in the conversion of substrate to product. Thus, in this case the K_m of the enzyme contributes very little to the carrying capacity of the enzyme; that is, as I will describe later, the height of a horizontal pipe on a vertical pipe little impacts on the diameter of that horizontal pipe and thus on the carrying capacity of that pipe (see Figure 4.1 for better clarity).

Secondary Data Plots

For a discussion on the underlying attributes of collecting enzymatic data and the use of such data in secondary data plots, for example, velocity versus substrate concentrations as in Figure 3.1 or a typical Lineweaver–Burk plot as in Figure 3.2, it is important to discuss some of the bases for obtaining such data plots that do not conform to those typically represented for enzymatic assay data.

For example, how might one interpret data illustrated by "open circles" in a modification of the data of possible data plots as illustrated in Figure 3.3. This modification of the possible data plots for a velocity versus substrate concentration where the plotted data are not curvilinear or where the measured velocity does not asymptotically approach some presumed V_{max} (Figure 3.3).

Both options in Figure 3.3 reveal measured velocity values (rates of an enzymatic reaction at the indicated substrate concentrations) that deviate from the normally described enzymatic reaction where the measured velocity asymptotically approaches a V_{max} value for that enzyme. In Option 1, the measured velocity appears to begin to plateau, but in the subsequent two data points at the highest substrate concentrations tested, there is a significant increase in a measured velocity. In Option 2, the measured velocities actually begin a steep decline in velocity at the two highest concentrations of substrate tested.

Let's examine how these two options might plot out in a Lineweaver–Burk data plot (Figure 3.4). In plotting the two open circle data points from Figure 3.3 into a typical Lineweaver–Burk data plot (Figure 3.4) we see that the normally straight line data plot for these kinds of data becomes a data line that turns upward (moving toward a smaller V_{max} value—Option 2) whereas the second data line (Option 1) turns downward moving toward a larger V_{max} value.

I have actually experienced (and obtained) Lineweaver–Burk data plots for enzymatic reactions similar to the two examples provided in Figures 3.3 and 3.4. The enzymatic reaction that provided for the upturn in the velocity values turned out to be two enzymes that each catalyzed the same conversion of

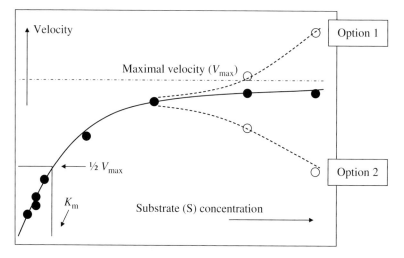

Figure 3.3 Typical velocity versus substrate concentration data plot showing the measured velocity asymptotically approaching maximal velocity at infinitely high concentrations of substrate (black data circles) and where the measured velocities do not asymptotically approach a maximal velocity, but rather turn upward (option 1) or downward (option 2) (open circle data points).

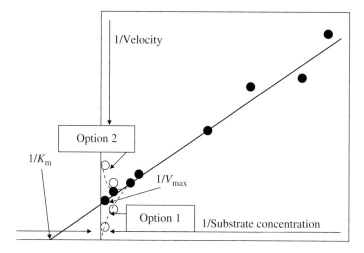

Figure 3.4 Typical Lineweaver–Burk data plot where the reciprocal of the velocity at the reciprocal of each substrate concentration is plotted. The values of V_{max} and K_m are calculated by extrapolation of where the data line crosses the *x* and *y* axes for 1/velocity and 1/substrate concentration (black circles) and for data plots that would be obtained from Options 1 and 2 from Figure 3.3 (open circles).

substrate to product. One of the enzymes had a smaller K_m than the other, but also a smaller V_{max} component. What I was actually measuring was two enzymatic reactions, and when the two enzymes were separated and assayed independently of each other it gave the two overlapping data plots provided in Figure 3.5. You should note that these current data plots are not derived from

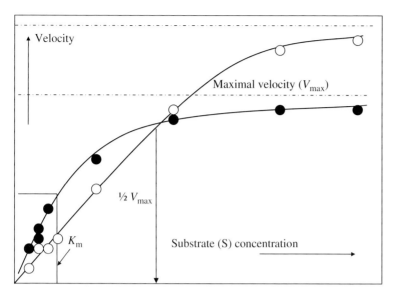

Figure 3.5 Typical velocity versus substrate concentration data plot showing the measured velocity asymptotically approaching maximal velocity at infinitely high concentrations of substrate (black data circles for one enzyme and open circles for a second enzyme).

real data (which is in my dissertation), but rather were generated as being representative of data you might obtain. Thus the actual velocity values from one data plot to the next data plot, although they may appear identical, are not intended to be identical or to represent some sum of two velocity values shown in some previous data plot.

I have yet to elucidate what contributed to the data illustrated in Figures 3.3 and 3.4 as Option 2. Possible explanations could be that the substrate, in addition to being the substrate of the enzymatic reaction, is also a noncompetitive inhibitor of the enzyme with a sufficiently low (small) K_{eq} value for binding at some second site on the enzyme that it is inhibitory only at the higher concentrations of substrate (which, by the way, could have been an explanation for the apparent activation of enzyme activity represented by Option 1?).

There are a great many ways in which one may obtain variable data for assessments of enzymatic activity. Indeed, I think the exceptions to the rules are data plots where one obtains the kinds of data illustrated in textbooks and in Figures 3.1 and 3.2 in this book. We generally assume that an enzyme is an enzyme irrespective of its concentration in some solution or cell extract. What kinetic parameters we might obtain in some initial cell extract may be vastly different once we have dialyzed the extract and/or concentrated the enzyme preparation by some means. To add insult to injury, we do not even know that the kinetic characteristics of an enzyme isolated from a cell are the same kinetic characteristics the enzyme has when it is functioning within a living cell—something researchers tend to forget or ignore too often. Enzymes are proteins, and proteins with enzymatic function tend to behave like most other proteins as they move from lower to higher concentrations—they aggregate. How might the aggregation of enzyme proteins change with respect to their kinetic characteristics when they function as monomers, versus when they function as dimers, trimers, or even aggregated to other proteins that may or may not possess enzymatic function?

We know, for example, that the enzymes in the pathway for aromatic amino acid biosynthesis function as a tightly linked aggregate where metabolic flow is controlled by the tight interactions between the substrate binding (catalytic) sites and metabolites are constrained within this pathway because the product of one enzyme is released in close proximity to the substrate-binding (and catalytic) site of the next enzyme in the metabolic pathway. If we isolate and purify these enzymes, can we expect them to have the same kinetic values as they have in their normal state within the living cell? Enzymes are curious molecules, and it is going to take curious researchers to learn to understand them. Do not go forward into advanced enzymology classes (or indeed any class) and accept the patent explanations for why the professor tells you something works (or does not work). Question any model of interactions between multiple elements in cellular metabolism. Such models were built and assembled based on interpretations of data, and those interpretations may be as faulty as the experiments upon which the interpretations were used to create a model.

This concludes your initial introduction to enzyme kinetics. It was very brief, very incomplete (as far as a good lecture on enzyme kinetics goes), and it tended to give you a biased perspective on two important enzume values (K_m and V_{max}) we will be using throughout this book. However, it now lets me introduce the concepts of pipe carrying capacity and height of feeder pipes in some water distribution system.

4

Metabolite Distribution Systems

Look up at the stars and not down at your feet. Try to make sense of what
you see, and wonder about what makes the universe exist. Be curious.

Stephen Hawking

This chapter will seek to explore how metabolites are distributed amongst differing metabolic pathways by attempting to relate the aforementioned enzyme kinetic values of K_m and V_{max} to a water distribution system such as one might see in a gravity-fed water distribution system in ancient Rome.

I'll wager that you never expected to be reading about water distribution systems in a book that implies it will teach you about enzymology. However, as was discussed in the previous chapter, I had begun to relate velocity of an enzymatic reaction to the diameter (or carrying capacity) of a water pipe, and the K_m of an enzymatic reaction to the relative height of a column of water in a gravity-fed water distribution system.

Let's first discuss the issue of K_m as it pertains to the height of a column of water. As you can see in Figure 4.1, as water flows into the system, it will spill over from the aqueduct and into what was probably an underground distribution system. However, because the water pressure due to the height of the aqueduct will be sufficient to force water up vertical standpipes, water can be delivered to individual horizontal pipes that serve to distribute the water into individual homes or public facilities. Those homes lowest on the vertical pipe will obviously be the first to receive water (and the last to stop receiving water during a drought). Homes higher up the vertical pipe will only receive water if the inflow of water to the system is greater than the outflow of water via the lower horizontal pipes. In addition, the amounts of water that will flow out of the vertical pipes via the horizontal pipes will correlate to the diameters of the horizontal pipes, and homes with the larger diameter pipes (presumably the homes of the rich and famous or royal) will receive more water than homes with the smaller diameter pipes (presumably the homes of the middle class). One assumes that the poor had to get their water from public wells, which had very small diameter pipes high up on the vertical distribution pipes.

You are probably ahead of me by now and understand that this hypothetical water distribution system is analogous to a metabolic pathway in a living cell. The K_m values of enzymes in such a metabolic pathway control the flow of

Enzyme Regulation in Metabolic Pathways, First Edition. Lloyd Wolfinbarger.
© 2017 John Wiley & Sons, Inc. Published 2017 by John Wiley & Sons, Inc.

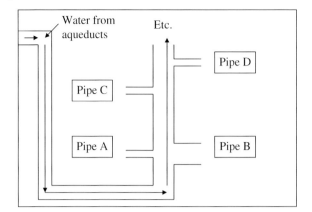

Figure 4.1 A diagrammatic illustration of a hypothetical water distribution system such as may have been in operation in ancient Rome. The water in this hypothetical system comes from some water source high in the mountains surrounding Rome and is delivered via an aqueduct into a water distribution system that feeds water to homes and public facilities in the city. Pipes A through D are intended to represent volumes of water that might be delivered based on two aspects of the distribution system: 1) the diameter, and thus carrying capacity, of the pipe feeding the home or public facility; and 2) the height of the pipes relative to other pipes in the water distribution system and thus the social/political stature of the owner of that home.

metabolites by requiring concentrations of substrates within that metabolic pathway to fall within about one log either side of the K_m value of that enzyme. As the concentration of a substrate for a given enzyme drops below the concentration equivalent to the K_m of that enzyme, the measured velocity of conversion of that substrate to product declines in a linear fashion, and where that substrate concentration is approximately one log lower in concentration than the K_m, the measured velocity will approximate to less than about 5% of the maximal velocity. As the concentration of substrate rises greater than the K_m value of an enzyme, the measured velocity initially increases in an approximate linear fashion until that concentration begins to approach a concentration approximately 10 times greater than the K_m value. At that point the enzyme is approaching saturation with substrate and the measured velocity is approaching the V_{max}. If you will visualize a column of water in the vertical pipe of Figure 4.1, and the water column is rising slowly, you will observe that water will start to flow through horizontal pipe B and then through horizontal pipe A. *For the moment, ignore the diameters of pipes A and B. The drawing was made to put the middles of pipes A and B at the same heights to give them the appearance of having the same K_m values (as enzymes) or the same middle of pipe relative to height on the vertical pipe. However, in order to do this, the diameter of pipe B had to be made larger than the diameter of the vertical pipe to illustrate that even if two enzymes possess the same K_m values, the enzyme with a greater V_{max} (i.e., larger diameter pipe) will have measurable conversion of substrate to product before the enzyme with the lower V_{max} (i.e., smaller diameter pipe).*

As the height of the column of water in the vertical pipe rises the volumes (rates or velocity of enzymatic reaction) of water flowing through the two pipes will increase until such time as the column of water in the vertical pipe is higher than the top inside diameter of each pipe (assumes that the horizontal pipes do not carry off more water than can enter the vertical pipe). At that time water flow through both pipes A and B will be maximal and the water will continue to rise in the vertical pipe until it reaches horizontal pipe C. When the water level in the vertical pipe is higher than the top inside diameter of horizontal pipe C, water flow through pipe C will be maximal and the water level will continue to rise (again assuming that the horizontal pipes do not carry off more water than can enter the vertical pipe) until it reaches horizontal pipe D. Eventually, according to the illustration, excess water will spill over the top of the vertical pipe (probably into a common (public) fountain or the sewer system).

If you can now shift to the idea that the water distribution system is analogous to a metabolic pathway, where each horizontal pipe represents an enzyme in that metabolic pathway, you can begin to understand that "feeding" substrate into the first enzyme in a metabolic pathway will result in the formation of product. Product in this instance is the water in the water vertical pipe moving into a "home." Which metabolic pathway that product enters is dictated by the K_m value of each of the enzymes (four in the hypothetical Figure 4.2). Those enzymes (enzymes A and B in this example) with K_m values around 1×10^{-8} (pipes A and B) will be able to bind substrate and convert it to some product at much lower concentrations of substrate than those enzymes with K_m values around 1×10^{-6}. If the V_{max} values of these enzymes (pipes A and B) are sufficiently large, that is, the carrying capacity of inside diameter of the pipes as representative of velocities of the enzymatic reaction, enzymes A and B will consume all of the substrate entering the metabolic system and no substrate will find its way into metabolic pathways where enzymes C and D (Figure 4.2, or pipes C and D according to the

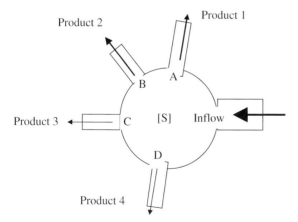

Figure 4.2 Cartoon to illustrate the role of enzymes (i.e., enzymes A, B, C, and D corresponding to horizontal pipes A, B, C, and D in Figure 4.1) in mediating the flow of substrate entering into a common "compartment" within the intermediary metabolism of a cell. The respective K_m and V_{max} values of each enzyme dictate which metabolic pathway receives metabolite and hence makes its specific product.

analogy in Figure 4.1) represent additional pathways for substrate entering into these multiple metabolic pathways. In Figure 4.2, the large arrow marked as "inflow" represents some commitment enzyme step into this multiple metabolic pathway system.

Carrying the analogy of Figure 4.2 to Figure 4.3, it becomes possible to describe a scenario where, as a result of cellular activity or the activity of some external "force" (e.g., erythrose-4-phosphate and phosphoenolpyruvate is poured into a flask of bacterial cells), a given metabolite becomes available as a substrate and three enzymes (i.e., in Figure 4.3) are available to compete for that substrate. If we use these metabolites poured into a flask of bacterial cells as a point for discussion, we might stipulate that the "Inflow" part of Figure 4.3 represents the transport mechanism whereby the metabolites are transported into the bacterial cell(s). In this case, the metabolites might become substrate by being transported, that is, they are chemically unchanged, and the cell is capable of utilizing three (four in Figure 4.2) different enzymes to convert these two metabolites into one or more "products." To illustrate this metabolite conversion into one or more products, we can turn to Figure 4.3.

The metabolic pathway illustrated in Figure 4.3 is consistent with the previously discussed examples of a water distribution system in ancient Rome (Figure 4.1) and competition for substrate by four hypothetical enzymes (Figure 4.2): Figure 4.3 illustrates compounds that serve as substrates, namely substrates 1 and 2, for one or more enzymes, that is, enzymes a and b, and c, respectively. This particular metabolic pathway eventually channels two substrate molecules (substrates 1 and 2) into three products (P_1, P_2, and P_3); which particular products are synthesized is controlled by the respective K_m and V_{max} values for those enzymes (e and f) that compete for a common substrate (Y) and then by enzymes g, h, and I, which metabolize substrates A and Z. As illustrated, the metabolic pathway in Figure 4.3 shows three enzymes (enzymes a, b, and c) as catalyzing the same substrate molecules (substrates 1 and 2) into the same product (compound X).

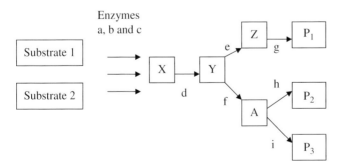

Figure 4.3 A diagram illustrating a metabolic pathway whereby two substrate molecules are metabolized to compound X via enzymes a, b, and c. Compound X is metabolized to compound Y by enzyme d. Compound Y is metabolized to compounds Z and A by enzymes e and f, respectively. Compound Z is metabolized to product P_1 by enzyme g, and compound A is metabolized to products P_2 and P_3 by enzymes h and i, respectively.

The metabolic pathway illustrated in Figure 4.3 was not chosen at random, but in fact illustrates the conversion of erythrose-4-phosphate and phosphoenolpyruvate into shikimate (via intermediary metabolites 3-deoxyarabinoheptulosonate-7-phosphate, 5-dehydroquinate, and 5-dehydroshikimate). The actual metabolic pathway has been shortened to protect the author from the reader. Shikimate (compound X in the diagram) is converted to chorismate (compound Y in the diagram) via an intermediary metabolite 3-enolpyruvylshikimate-5-phosphate. Chorismate is converted into two different compounds, prephenate and anthranilate (compounds A and Z, respectively), via enzymes f and e as illustrated in Figure 4.3. Prephenate (compound A) is converted to phenylpyruvate (and finally to phenylalanine) and to para-hydroxyphenylpyruvate (and finally to tyrosine) via enzymes h and i. Anthranilate is finally converted into the amino acid tryptophan (Product 1), phenylpyruvate is converted into the amino acid phenylalanine (Product 2), and para-hydroxyphenylpyruvate is converted into tyrosine (Product 3). It might seem like cruel and unusual punishment to have gone into this level of detail in a metabolic pathway. However, it was necessary as this pathway will be used in future chapters to illustrate the role of K_m and V_{max} values of enzymes in controlling the flow of metabolites through metabolic pathways. Figure 4.4 is the same as Figure 4.3 except I have added the actual intermediary metabolites in this pathway. I have shortened the pathway (omitted several intermediates) to protect the innocent; if you want the full metabolic pathway I would refer you to a standard biochemistry textbook.

The pathway for aromatic amino acid biosynthesis was chosen specifically because it is very complicated and because it is highly regulated. The mechanisms of regulation involve modifiers binding to enzymes at critical points in the pathway to alter the kinetics of individual enzymes in order to modulate the flow of metabolites into one or more of the three aromatic amino acids, depending on the availability and/or need for a particular aromatic amino acid. In subsequent

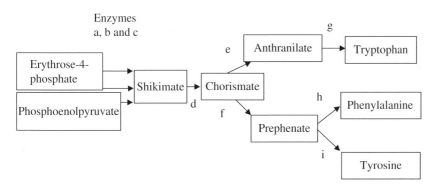

Figure 4.4 Pathway as shown in Figure 4.3, but with specific molecules in this aromatic amino acid biosynthetic pathway substituted for the symbols used in Figure 4.3. Prephenate is actually metabolized to either phenylpyruvate prior to formation of phenylalanine, or to parahydroxyphenylpyruvate prior to formation of tyrosine, but these intermediates were left out of this figure in order to emphasize the enzyme steps relevant (important) to regulation of this metabolic pathway and to aid in comprehension.

figures I propose to revert to the symbols used in Figure 4.3 for the actual metabolic intermediates (used in Figure 4.4) in order to emphasize the regulatory events. If you find it difficult to follow, please refresh yourself by referring to Figure 4.4 as I discuss Figures 4.5, 4.6, and 4.7.

Figures 4.5, 4.6, and 4.7 will be utilized to illustrate some of the complex regulatory aspects of aromatic amino acid metabolism. A "Pos" in a box will indicate a positive regulatory event (increasing the velocity of conversion of substrate to product), and a "Neg" in a box will indicate a negative regulatory event (decreasing the velocity of conversion of substrate to product). You will be easily confused by Figures 4.5, 4.6, and 4.7 unless you dissect one event at a time until you can "see" how each regulatory event shifts the direction of flow of metabolites through the entire metabolic pathway. Further, when I describe some regulatory event as

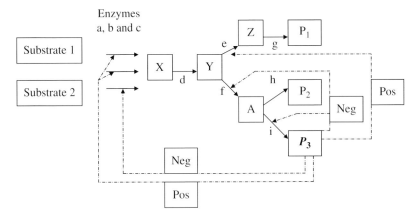

Figure 4.5 Diagram of the aromatic amino acid biosynthetic pathway showing the enzymatic steps in the pathway at which tyrosine (P_3) modifies enzymatic activity via positive and negative regulatory events.

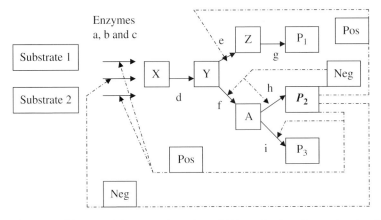

Figure 4.6 Diagram of the aromatic amino acid biosynthetic pathway showing the enzymatic steps in the pathway at which phenylalanine (P_2) modifies enzymatic activity via positive and negative regulatory events.

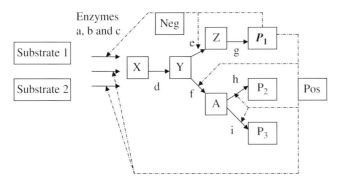

Figure 4.7 Diagram of the aromatic amino acid biosynthetic pathway showing the enzymatic steps in the pathway at which tryptophan (P$_1$) modifies enzymatic activity via positive and negative regulatory events.

increasing or decreasing the velocity of conversion of substrate to product, there is no intent to imply that this increase or decrease is mediated via some change in the apparent K_m or apparent V_{max}, only that the overall rate of conversion of substrate to product at some specific enzyme-mediated event increases or decreases with regulation.

For example, there is some simplicity to the regulatory events. Tryptophan basically activates those enzymes that lead to the synthesis of tyrosine and phenylalanine and inhibits those enzymes that lead to the synthesis of tryptophan. Similarly, phenylalanine activates those enzymes that lead to the synthesis of tryptophan and pretty much inhibits those enzymes that lead to the synthesis of phenylalanine (and a bit for the synthesis of tyrosine). Tyrosine, analogous to phenylalanine, activates those enzymes that lead to the synthesis of tryptophan, but inhibits those enzymes that lead to the synthesis of tyrosine (and a bit for the synthesis of phenylalanine). The three isozymes that synthesize 3-deoxyarabinoheptulosonate-7-phosphate (leading to shikimate) from erythrose-4-phosphate and phosphoenolpyruvate are each inhibited by one of the end products of aromatic amino acid biosynthesis and activated by two of the end products of aromatic amino acid biosynthesis. It is logical that if one of the three aromatic amino acids is abundant, there is less need to synthesize early and common precursor metabolites in the aromatic amino acid biosynthetic pathway, and this can easily be accomplished by shutting down one of the three isozymes forming the gatekeeping enzyme for the whole pathway and activating those enzymes that lead to the synthesis of the less abundant aromatic amino acids.

Other metabolic pathways in intermediary metabolism are regulated by similar means. However, such regulation can also occur by controlling the levels of enzymes transcribed and synthesized, and such regulation can be accomplished by an "excess" of a given end product of some metabolic pathway. Because here the focus of regulation is on the enzyme, the regulation of transcriptional and or translational events in enzyme synthesis is not covered in this text.

Cells also compartmentalize metabolic pathways and the flow of metabolites via multiple mechanisms. Some enzymes are compartmentalized within subcellular organelles such as mitochondria, nucleus, Golgi bodies, and so forth, and the "gatekeeper" enzyme for some metabolic pathway may be more of a transport mechanism whereby the substrate is sequestered within some subcellular organelle and thus unavailable to enzymes that could metabolize that substrate were it available in the general cytoplasm. Some enzymes are clustered into an enzyme aggregate whereby one enzyme binds substrate, converts it to product and the next enzyme gains priority access to that "product" (substrate for the second enzyme) because its substrate is released from the catalytic site of the preceding enzyme in the immediate vicinity of the catalytic/binding site of the second enzyme. This compartmentalization mechanism controls where metabolite goes by fixing the enzymes in some metabolic pathway in close proximity such that although their "relative affinity" (K_m) for a given substrate molecule may be lower than other enzymes, they will have priority access to that substrate by virtue of physical proximity.

Such mechanisms for controlling the flow of metabolites through metabolic pathways are not, however, the topic of this book. Based on information in Chapters 1 and 2, you should now have some basic understanding of K_m and V_{max} values for individual enzymes, of how K_m and V_{max} values are used in intermediary metabolism to direct the flow of metabolites into different metabolic pathways, and that metabolic pathways can be regulated at the level of the enzyme by metabolites specific to that metabolic pathway.

Thus, the next question to be considered is how does a metabolic intermediate modify the enzymatic function of an enzyme positively, negatively, or both?

Let's think about the enzyme 3-deoxyarabinoheptulosonate-7-phosphate synthetase. As you remember from Figure 4.4 there are three isozyme forms of this enzyme that convert phosphoenolpyruvate and erythrose-4-phosphate to "shikimate" (remember I shortened the pathway in Figure 4.4, and 3-deoxyarabinoheptulosonate-7-phosphate is actually the product of the synthetase). In subsequent figures (Figures 4.5–4.7) I illustrated that each isozyme performed the same catalytic reaction and that each isozyme was regulated by each of the end products (phenylalanine, tyrosine, and tryptophan) of the aromatic amino acid biosynthetic pathway. In that I could use any one of the three isozymes (enzymes a, b, and c in the figures) to make the same point, it is prudent for me to choose to discuss only one isozyme at a time. Consider the options for such a complicated regulatory system. There, for the sake of simplicity, should be one catalytic site and perhaps three "independent" (?) regulatory sites on the enzyme. By the use of "independent" regulatory sites, we should start with the most simple option where phenylalanine binds at one regulatory site and achieves its regulatory activity, tyrosine binds at one regulatory site and achieves its regulatory activity, and tryptophan binds at one regulatory site and achieves its regulatory activity. This option is visualized in Figure 4.8.

We can use the same figure to illustrate each of the three synthetase isozymes (enzymes a, b, and c in Figures 4.5–4.7), needing only to change whether a particular end product exerted a positive (activation) or negative (inhibition) effect on the isozyme. For purposes of this discussion, let's consider that the isozyme

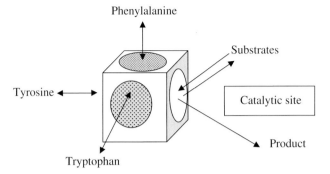

Figure 4.8 Cartoon illustrating three binding sites for end-product regulation and one catalytic site for the enzyme 3-deoxyarabinoheptulosonate-7-phosphate synthetase.

we want to discuss is feedback inhibited by phenylalanine and feedback activated by tyrosine and tryptophan (refer back to Figure 4.6).

In trying to create equations in the manner that I think you should begin to get used to, I propose to describe regulatory options in two stages. In the first stage (see Equation 4.1), I will let phenylalanine be an "activator" and tyrosine be an "inhibitor" so that we can talk about how the rate constants might change for the nonmodified and modified enzymatic mechanisms for the conversion of phosphoenolpyruvate and erthrose-4-phosphate to "shikimate." In the mechanisms in Equation 4.1, I propose that we continue to operate under quasi-equilibrium assumptions, but I also propose to utilize a different mode for activation and/or inhibition, which I will be using in later chapters, so pay attention to the rate constants and to the discussion of why some change to a rate constant will achieve the inhibition or activation of flow of metabolites through the overall pathway to end product(s).

The first set of equations to define mechanisms for modifying an enzyme are shown in Equation 4.1. This is the first time you are seeing such an approach to defining enzymatic mechanisms and I would encourage you to try to follow the individual steps and associated rate constants rather than trying to deal with the whole of the mechanisms. In addition, there are "ground rules" to this approach. One ground rule is that in the presence of a modifier that is binding to an enzyme, you should understand that the modifier is infinite in amount and always in excess of the enzyme it is modifying, so that binding of the modifier to the enzyme does not change the concentration of modifier. A second ground rule is that the substrate(s) are always present in infinite amounts and at saturating concentration levels so that each mechanism achieves a maximal and constant concentration of the enzyme–substrate(s) complex, whether the enzyme–substrate(s) complex is modified or unmodified. In reality, all I am trying to convey to you is that each mechanism will be converting substrate(s) to product at some maximal velocity and that any modifier will divert as much free enzyme to modified enzyme as the association/dissociation constants defined by the described rate constants will permit.

Again, pay attention to the relative values assigned to specific rate constants and work back and forth between the Equation (the modified and nonmodified enzyme mechanisms) and the discussion. I will be repetitive in my explanation to try to help, but you will need to work hard to follow this first introduction to my teaching mode as I will use it repeatedly in subsequent chapters.

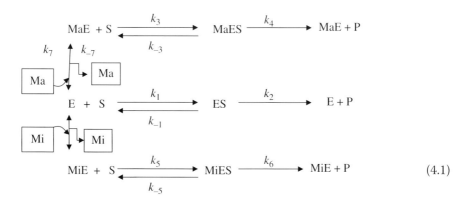

$$(4.1)$$

So, what is Equation 4.1? If you will examine the diagram carefully, you will see a middle enzyme mechanism that is the same as the mechanism you saw in Equation 3.2. This "middle enzyme mechanism" represents the conversion of substrate(s) (in this case phosphoenolpyruvate AND erythrose-4-phosphate) to product ("shikimate") via some nonmodified enzyme mechanism. In this "middle enzyme mechanism," k_2 is much much ($<<$) less than both k_1 and k_{-1} (and k_1 is equal to or greater than k_{-1}). We write these relationships as $k_2 << k_1 \geq k_{-1}$. This relationship of the rate constants in the nonmodified mechanism means we are working under quasi-equilibrium assumptions and $V_{max} = k_2[ES]$ and $K_m = k_{-1}/k_1$. Remember, K_m for this mechanism is actually $k_{-1} + k_2/k_1$, but because $k_2 << k_1$, k_2 in the K_m value is negligible, K_m can be considered to be k_{-1}/k_1.

In the absence of modifier (modification of enzymatic activity), all substrate(s) will be converted to product via the "middle enzymatic mechanism."

However, let's look next at the "top enzyme mechanism." You can see that in the presence of modifier (Ma, where the "a" implies activation), the modifier (tyrosine) binds reversibly to free enzyme resulting in some equilibrium concentration of free enzyme (E) and modified enzyme (MaE). Modified enzyme (MaE) is capable of reversibly binding to substrate (S) forming an modified enzyme–substrate complex (MaES), and the MaES complex can break down to either product (P) and modified enzyme (MaE), or to substrate (S) and modified enzyme (MaE). Remember that we are under quasi-equilibrium assumptions, so we can say that $k_4 << k_{-3} \leq k_3$. Once again, this would mean that the V_{max} for the modified (activated) enzyme mechanism is $V_{max} = k_4(MaES)$ and $K_m = (k_{-3} + k_4)/k_3$, and since k_4 is insignificant with respect to either k_{-3} or k_3, we can rewrite K_m for this mechanism as $K_m = k_{-3}/k_3$.

In the presence of modifier Ma, we now have two mechanisms for the conversion of substrate to product, the nonmodified mechanism and the modified mechanism.

Now, ordinarily I would simply jump to describing the "bottom mechanism" where the inhibitory modifier (Mi, where the "i" implies inhibition) is present, but since this is your first effort with this concept, I propose to repeat this whole concept in a second discussion to follow this discussion of a modifier being an activator of the enzyme conversion of substrate to product.

First, let's set parameters for the binding of the modifier tyrosine (Ma). We do this by establishing rate constants for the binding and release (unbinding) of the modifier to the free enzyme. Since we show this step as reversible, we will have two rate constants that we will define as k_7 for the association of the activator (Ma) (tyrosine) to free enzyme and k_{-7} as the dissociation of the activator (Ma) (tyrosine) from the modifier–enzyme complex to yield free enzyme and tyrosine. If we stipulate that $k_7 >> k_{-7}$, we will have, for all intents and purposes a "nonreversible binding" of modifier to free enzyme since the dissociation of modifier will be negligible relative to the binding of modifier. Let's not do this. Let's set $k_7 = k_{-7}$ such that in so doing we will have an approximate equal distribution of free enzyme (E) and modified enzyme (MaE) to bind with substrate (S) (again, remember "substrate" is really both phosphoenolpyruvate AND erythrose-4-phosphate), so we will have two mechanisms operating to convert substrate to product once we attain some kind of equilibrium in the binding of modifier to the enzyme.

Work your way through this situation. If you let $k_1 = k_3$, $k_{-1} = k_{-3}$, and $k_2 = k_4$, the modified mechanism will present the same apparent K_m and apparent V_{max} values as the nonmodified mechanism, and you will not observe any change in the kinetics of the conversion of substrate to product.

If you let $k_1 = k_3$, and $k_{-1} = k_{-3}$, but let k_4 be greater than k_2 (i.e., $k_4 > k_2$), the modifier will not change the apparent K_m for the enzymatic conversion of sub-strate to product, but the modifier will change the apparent V_{max} value for the conversion of substrate to product. The new apparent V_{max} value in the presence of tyrosine (modifier) will be greater, and although both the nonmodified and modified mechanisms are converting substrate to product, you will still observe an apparent greater rate of conversion of substrate to product and the modifier will appear as an activator in your data plots.

If you let $k_1 < k_3$ (which is just another way of saying k_3 is greater than k_1), let $k_{-1} = k_{-3}$, and let $k_2 = k_4$, what will you have in the way of a new change in the rate of conversion of substrate to product in the presence of modifier (tyros-ine)? Think about this situation. Remember, k_7 still equals k_{-7} ($k_7 = k_{-7}$), so we will still have an equal amount of free enzyme concentration and modifier-modified free enzyme concentration when some equilibrium has formed, but since k_3 is greater than k_1, more MaE will bind with substrate than E will bind with substrate. All other things being equal, when the system has achieved some equilibrium state there will be greater amounts of MaES than ES, and given that the rates of conversion of substrate to product for each of the two operative mechanisms will approximate $v = k_2(ES)$ AND $v = k_3(MaES)$, you should observe a subtle increase in the flow of substrate to product through the modified mechanism. Although we began with "saturating" concentrations of substrate for a K_m value equalling k_{-1}/k_1, the K_m value for the modified enzyme is now k_{-3}/k_3 and this will make the saturating concentration of substrate even lower for the modified mechanism. For example, if we stipulate the old K_m for the nonmodified mechanism was 1×10^{-6} M and for the modified mechanism was 1×10^{-7}, we would have a concentration of substrate that was 10 times greater than the concentration needed to "saturate" the nonmodified enzyme. Since we asymptotically approach V_{max}, there is always a bit of extra velocity to be achieved with a supersaturated enzymatic reaction. Admittedly, this approach to achieving "activation" of the enzymatic reaction will yield small activation values; I wanted you to think about taking some approach other than increasing the K_{cat} (k_4 in this instance) rate constant, which would be more obvious as a means of activating the enzymatic reaction. This is where you should also realize that you can substitute tryptophan for tyrosine in this discussion since according to Figure 4.8, both tryptophan and tyrosine negatively feedback inhibit this particular isozyme. Question: If you make k_7 just greater than k_{-7}, would shifting more of the enzyme in the system toward the modified mechanism make the differences in the conversion of substrate to product via the modified mechanism more "obvious" as an activation of the conversion of substrate to product? Later on, in Chapter 5, you will learn about whether or not what we just arrived at was a competitive, a noncompetitive, or an uncompetitive modifier, but for now just let's try another approach to regulation of this isozyme.

Let's think about phenylalanine inhibiting this enzyme when phenylalanine is abundant in the cell and there is no need to be synthesizing more phenylalanine. Obviously you need to refer back to Figure 4.8 and Equation 4.1 to get your bearings, but now focus on the "bottom" mechanism in Equation 4.1. I know that at this point you are anxious to get to the situation where all three isozymes are being active and regulated by differing concentrations (abundance of or lack of the three end-product reactants: phenylalanine, tyrosine, and tryptophan) and you want to model mechanisms for all possible contingencies, but let's try to approach things as simply as possible and one at a time.

Let's repeat Equation 4.1 so you don't have to keep flipping back and forth. As with the previous discussion, we still have the nonmodified enzyme mechanism as the "middle enzyme mechanism." The "bottom enzyme mechanism" becomes relevant when modifier "Mi" is introduced into the system. In this case, we understand that "Mi" is the end product phenylalanine and it is appropriate that phenylalanine be able to reduce the conversion of phosphoenolpyruvate and erythrose-4-phosphate to "shikimate" (and ultimately to phenylalanine)—if you can get it elsewhere, you don't need to synthesize it. In this instance, the modifier (phenylalanine, or "Mi") will bind to free enzyme E resulting in the MiE complex. According to the "bottom enzyme mechanism," the modified free enzyme (MiE) will bind to substrate (S) (which you remember is actually two compounds) to form a MiES complex. Formation of this MiES complex is reversible, and some relationship of k_5 and k_{-5} (plus k_6, which is negligible in the K_m value) will be involved in the K_m of this modified enzyme mechanism. The V_{max} of this modified enzyme mechanism can be defined by $V_{max} = k_6([\text{MiES}])$, and I will remind you of the "ground rules" described earlier about concentrations and infinite amounts of substrate and modifiers. We are also under quasi-equilibrium assumptions for this modified mechanism, so $k_5 \geq k_{-5} >> k_6$.

We are once again short of rate constants for the binding of modifier "i" to free enzyme and for the dissociation of modifier from the MiE complex to yield free E and free Mi. So, let's repeat Equation 4.1 below:

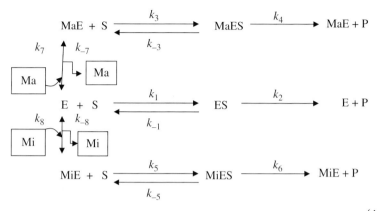

(4.1 repeated)

You will note that Equation 4.1 now has two rate constants for the interaction of modifier "Mi" (phenylalanine) with free enzyme. Rate constant k_8 relates to the association of the modifier Mi with free enzyme E, and the rate constant k_{-8} relates to the dissociation of the modifier Mi from the MiE complex to re-form free enzyme (E) and free modifier (Mi). Rather than go through all possible relationships of these two rate constants, let's just stipulate that $k_8 > k_{-8}$, meaning that in the presence of an abundance of modifier this enzyme mechanism will come to some equilibrium between the concentrations of enzyme ([E]) and modified enzyme ([MiE]). With k_8 being larger than k_{-8}, we can further stipulate that k_8 is 10 times greater than k_{-8} and save on discussion. This would mean that at equilibrium, there will be an approximately 10 times greater concentration of MiE than E. We thus favor the equilibrium toward the "bottom enzyme mechanism." If we set $k_{-8} > k_8$ and stipulate that k_{-8} is 10 times greater than k_8, we have just the opposite of what we stipulated previously and we would now favor the equilibrium toward the "middle enzyme mechanism"—the nonmodified mechanism.

So, how do you want to set the relationships of k_1 to k_5, k_{-1} to k_{-5}, and k_2 to k_6, and which "enzyme mechanism" (middle or bottom) do you want to favor to work under? In that I think you probably have gotten the idea by now, I would prefer to leave you to your own devices and move on to the next discussion.

Now, can you imagine the situation where all three end products of the aromatic amino acid biosynthetic pathway are present to varying concentrations in the cell? Each isozyme is presumed to be regulated by all three end-product amino acids, some in a positive manner and others in a negative manner. You might ask why is this relevant? If this is a "commitment step" in the aromatic amino acid biosynthetic pathway, why would it be important that two of the end-product amino acids activate (provide a positive regulatory event) a given isozyme. Why not just enable each end-product amino acid to inhibit its specific isozyme (why not just have one modifier binding site on each isozyme and let the modifier be an inhibitor?)? That way, if only one end-product amino acid were in abundance in the cell, only one of the three isozymes would be shut down in enzymatic activity. If all three end-product amino acids were in abundance in the cell, all three isozymes would be shut down and aromatic amino acid synthesis would be minimal (or "zero"?). The answer lies in the fact that metabolic intermediates such as shikimate (as one example here) are used in other metabolic pathways and thus a living cell that has slowly evolved multiple metabolic capabilities survives best if no one regulatory event totally controls a given enzymatic reaction. So, this is where you go back to Figures 4.5–4.7 and look which enzymes between "shikimate" and the end-product amino acids regulate which enzymes in the different pathways, and whether that regulation is "positive" or "negative." What you will observe, before you go blind, is that each individual end-product amino acid will act not only in a negative manner on enzymes in its specific pathway, but also in a positive manner on enzymes not in its specific pathway. In this way the flow of necessary intermediates is not limited to metabolic pathways where some end product is in a "low concentration" in the cell. Since each intermediate in the aromatic amino acid biosynthetic pathway is also used elsewhere, no specific enzyme is completely inhibited—presumably due to the presence of multiple modifier binding sites on each enzyme in intermediary metabolism.

So how are we ever to figure out such complexity in the regulation of a given enzyme, let alone figure out if that regulation modulates the K_m, the V_{max}, or both the K_m and V_{max} of a given enzyme mechanism. "We" won't, but you can. When I took enzymology in the late 1960s and early 1970s I was lucky to be able to manually plot my data on graph paper and draw my figures for my dissertation using a Leroy lettering kit (a Leroy lettering kit contained plastic flat bars with letters cut into the plastic, a three-legged drawing pen where one leg fits into the numbers/letters on the flat bars and one leg rides along in a long groove next to the numbers/letters, and the third leg is an ink pen.) Do you realize how great it is now to have computer programming that will let you enter specific values for all these rate constants plus concentrations of enzyme and substrate and let the computer plot your data for you? By the way, I forgot to add to my story above that when I took enzymology all I had was a hand calculator (actually a "slide rule" if you know what that is) AND I had to walk to school uphill in both directions (and through the snow with no shoes). Do you also understand, and this is why this book, that unless you realize that you can make up your own mechanisms and values for rate constants, that you can figure out how the enzyme you study may be actually converting substrate to product, and what agents may modify that enzyme in what manner, that I have failed you? You have to be able to work things out in your mind before you will know how to use the computer programming to answer your questions.

Enough Philosophy 101. Let's go further along in this characterization of enzyme mechanisms and how we can determine from our plotted data what kinds of options we have with respect to enzyme mechanisms and regulatory events.

5

Modification of Enzymatic Activity

*How does a fly follow a trail? Close observation reveals that he tends to walk
in more or less straight lines until something diverts him.*
Vincent G. Dethier, *To Know a Fly* (Holden-Day)

You were earlier introduced to the concept of a substrate-binding site whereby a substrate molecule binds to the "surface" of an enzyme on its way to being converted to product. In the rather simplistic quasi-equilibrium assumption model given to you, it was possible to illustrate the roles of rate constants (for example k_1, k_{-1}, and k_2) to illustrate the difference between an affinity constant (K_{eq}) and a Michaelis constant (K_m). In Chapter 4 the concept of "positive" and "negative" modification of an enzyme by an "end product" of a specific metabolic pathway was presented to you.

This chapter will attempt to explain means of regulating enzymatic activity by modifying either the apparent K_m and/or the apparent V_{max} of an enzyme. I use the word apparent here because the means of regulation to be described to you are not actually modifying the "real" K_m and/or V_{max} values as we might be able to determine in some unmodified state. Rather, what will be described will be the shifting of the conversion of substrate to product via alternative mechanistic pathways. New K_m and V_{max} values will present themselves; however, the original K_m and V_{max} values of the unmodified enzymatic mechanism remain unchanged. This shifting is only one means of regulating the activity of an enzyme at the level of the enzyme, but for purposes of the present discussion I will use it because I think it will help you to better understand other mechanisms that you will learn of in higher level classes in enzymology.

As a way of describing these means of regulating the K_m and/or V_{max} of an enzyme by some modifier, I will need to develop more fully the concept of a modifier-binding site. This modifier-binding site will be some area on the "surface" of the enzyme, very similar to the substrate-binding site in the orientation of specific hydrophobic and/or hydrophilic regions, where a modifier molecule may bind. In so binding, this modified form of the enzyme will possess different catalytic properties as compared to the nonmodified form of the enzyme. There may be multiple modifier-binding sites on an enzyme, and thus an enzyme can be subject to complex and multiple mechanisms for the conversion of substrate to product, as discussed in Chapter 4. In the examples provided in this chapter, I will restrict you to what amounts to "single" modifier-binding sites on an enzyme, but

I think with practice you can begin to consider how two modifiers (perhaps one acting in a "positive" manner and one acting in a "negative" manner) might permit creation of new regulated pathways.

Biochemistry teaches that enzymes are modified via one or more of three "common" mechanisms: 1) competitive, 2) noncompetitive, and 3) uncompetitive. Differing terminologies have been applied to modification of enzymes over the years by differing authors; however, for the most part, whatever one chooses to call the modification of an enzymatic event they contain common features. In order to maintain some connections with tradition, we will consider modifiers as being one or more agents that diverts the fly "from his/her straight line."

However, before I get into modifiers as either "inhibitors" or "activators" I need you to consider whether a modifier changes (increases or decreases) the measure of velocities of the enzymatic reaction (i.e., changes the apparent V_{max}) or whether a modifier changes (increases or decreases the concentration of substrate necessary to achieving some measure of velocities of the enzymatic reaction) the apparent K_m. As you will see, it is possible for a modifier to increase the apparent K_m of an enzyme reaction, but for the overall V_{max} value to increase and for the overall effect of the modifier to be an "activator." Similarly, it is possible for a modifier to decrease the apparent K_m of an enzymatic reaction, but for the overall V_{max} value to decrease and for the overall effect of the modifier to be an "inhibitor." It is important that you begin to think about what you mean (and others mean) when they call a modifier an inhibitor or an activator. It is possible for a modifier to be both an inhibitor and an activator (at the same time) depending on one's perspective at the time of analysis."

Competitive Modifiers

A competitive modifier is typically described as a solute that alters an enzymatic reaction by changing the apparent K_m, but not the apparent V_{max}, of the enzyme. Many competitive modifiers (inhibitors) are structural analogs of the substrate and may exert their modification (inhibition) by binding at the substrate-binding site, restricting or competing with the binding of substrate. However, I will describe alternative means in Chapter 6 by which competitive modifiers may modify an enzymatic reaction, to avoid your learning some mechanism by rote memory. I would prefer that you learn to understand and think about how you might better explain some observed enzymatic event than trying to make your observations fit into some model taught in most textbooks. Most instruction with respect to competitive modifiers describes the event as competitive inhibition (where the apparent K_m value increases—goes from negative 6 logs to negative 3 logs molar concentration, e.g.) and rarely makes provision for competitive activation (where the apparent K_m value decreases—goes from a negative 6 log to a negative 8 log molar concentration, e.g.). This process is best illustrated utilizing the Lineweaver–Burk data plot, to which you have already been introduced. A data plot illustrating competitive modification of an enzyme is shown in Figure 5.1.

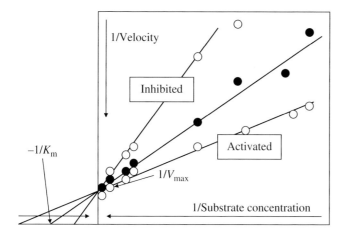

Figure 5.1 Lineweaver–Burk data plot showing a nonmodified enzymatic reaction (dark circles) and two modified enzymatic reactions (clear circles) where one modified reaction is classified as an inhibited enzymatic reaction (K_m value is increased) and the second modified reaction is classified as an activated enzymatic reaction (K_m value is decreased) relative to the K_m of the nonmodified enzymatic reaction.

As illustrated in Figure 5.1, addition of a modifier to the enzymatic reaction does not alter the intercept on the y axis and thus the V_{max} of a modified enzymatic reaction is unchanged relative to the nonmodified enzymatic reaction. A modifier that increases the measured K_m (meaning it changes the value from something like 1×10^{-6} M to something like 1×10^{-3} M), is said to be an inhibitor of the enzymatic reaction. A modifier that decreases the measured K_m (meaning it changes the value from something like 1×10^{-6} M to something like 1×10^{-8} M) is said to be an activator of the enzymatic reaction. An enzyme inhibited by a competitive modifier will require a higher substrate concentration than the nonmodified form of the enzyme to achieve a measured velocity equal to ½ the V_{max} if the modifier is actually competing with the substrate for binding to the substrate-binding site. An enzyme activated by a competitive modifier will require a lower substrate concentration than the nonmodified form of the enzyme to achieve a measured velocity equal to ½ the V_{max} and it is assumed in this form of competitive activation that the modifier is not competing with substrate for binding to the substrate-binding site. This would be a good time for you to fix in your mind how some intercept moving up or down, left or right in a double reciprocal data plot relates to velocities and/or substrate concentrations. I will try to repeat the essential elements of how the use of a double reciprocal data plot can be confusing throughout later chapters (where appropriate), but try to remember that for changes to K_m valuations, this data plot shows the reciprocal of K_m (negative one over K_m) and care must be taken to note that as K_m increases (concentration of necessary substrate increases) the negative one over K_m value actually decreases, and as K_m decreases (concentration of necessary substrate decreases) the negative one over K_m value actually increases.

We may illustrate how this modification might occur by using an equation similar to that shown in Equation 3.2, but adding two additional routes by which substrate may be converted to product by enzyme (E) (Equation 5.1). One route is where modifier Ma binds to free enzyme (E) to form a modified form of the enzyme, MaE. A second route is where modifier Mi binds to free enzyme (E) to form a modified form of the enzyme, MiE. For purposes of relating Equation 5.1 to Figure 5.1, the modifier Ma will be treated as being an activator of enzyme E, and the modifier Mi will be treated as being an inhibitor of enzyme E.

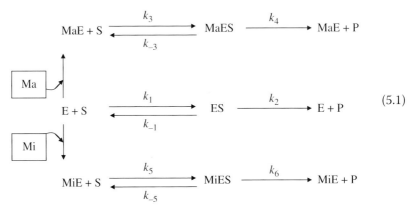

$$(5.1)$$

As illustrated, Equation 5.1 is incomplete and we need to further define the relationships of the rate constants. I must first, however, remind you that the binding of modifier to free enzyme is not reversible, as indicated by a unidirectional arrow in Equation 5.1. In that we have been operating under quasi-equilibrium assumptions for the sake of simplicity, let's continue to operate under quasi-equilibrium assumptions. We can do this by setting specific relationships for those rate constants within each of the three mechanisms illustrated in Equation 5.1. These relationships will be as follows:

k_1 is equal to or greater than k_{-1} and both are much much greater than k_2, which is more easily written as:

$$k_1 \geq k_{-1} \gg k_2$$

Similarly:

$$k_3 \geq k_{-3} \gg k_4 \text{ and } k_5 \geq k_{-5} \gg k_6$$

If we now set k_2 equal to k_4 and k_6 (i.e., $k_2 = k_4 = k_6$), we are stipulating that the rate-limiting steps for all three of these mechanisms (the nonmodified, the activated, and the inhibited enzymatic reactions) are identical and thus we are basically saying that neither modifier is modifying the $V_{\max}$ values for the conversion of substrate to product.

Remember, $V_{\max} = k_2[\text{ES}]$ for the nonmodified mechanism, $V_{\max} = k_4[\text{MaES}]$ for the modified (activated) mechanism, and $V_{\max} = k_6[\text{MiES}]$ for the modified (inhibited) mechanism, and under the "usual" conditions where a modifier is in excess and at high concentration, only one mechanism at any given option will be converting substrate to product.

To change the K_m values for the modified enzymatic mechanisms (as described, a prerequisite for a competitive modifier) all that we need to do is state that k_1 does not equal either k_3 or k_5. However, we probably should also specify that k_{-1} equals k_{-3} and k_{-5} just to keep things simple. If we state that k_3 is greater than k_1 we are changing (increasing) the rate constant for association of the substrate (S) to the modified enzyme MaE—but not changing the rate constants for the dissociation of substrate (S) from the modified enzyme (MaE)—such that substrate will bind more rapidly at each concentration of substrate to MaE than to E and at a given concentration of substrate (S), and thus more MaE will be converted to MaES than E would have been converted to ES in the absence of modifier at each substrate concentration except at saturating levels of substrate. We have activated (increased) the rate of conversion of substrate to product at nonsaturating concentrations of substrate by making more MaES available at each concentration of substrate (S) than would have been available to a nonmodified form of the enzyme. By increasing the concentration of the MaES complex at each concentration of substrate, the rate of conversion of substrate to product is increased.

Similarly, if we state that k_5 is less than k_1, we are changing the rate constant for the association of the substrate (S) to the modified enzyme MiE—but not changing the rate constants for the dissociation of substrate (S) from the modified enzyme (MiES)—such that substrate will bind more slowly to MiE than to E and at a given concentration of substrate (S), less MiE will be converted to MiES than E would have been to ES in the absence of modifier. We have inhibited (decreased) the rate of conversion of substrate to product at all sub-saturating concentrations of substrate by making less MiES available at each concentration of substrate (S) than would have been available to a nonmodified form of the enzyme. By decreasing the concentration of the MiES complex at each concentration of substrate except for saturating concentrations, the rate of conversion of substrate to product is decreased.

It should be obvious that what I have described is only one means for modifying the enzyme in a competitive modifier mode. In the example shown in Equation 5.1 and subsequent further defining of the rate constant relationships, we could easily complicate the modification mechanism by making the binding of modifier to free enzyme reversible. However, we would then have a much more complicated mechanism to explain and one that would be beyond the desired scope of this example of a competitive modifier.

Noncompetitive Modifiers

A noncompetitive modifier is typically described as a solute that changes the observed V_{max} of an enzymatic reaction without changing the calculated (apparent) K_m of that enzymatic reaction. Once again, this means of modification of an enzymatic reaction is easily illustrated using the Lineweaver–Burk data plot (Figure 5.2).

As illustrated in Figure 5.2, a noncompetitive modifier does not alter the calculated K_m value relative to the nonmodified enzyme (all three data plots intercept at the same point on the *x* axis). However, the calculated V_{max} values for both the

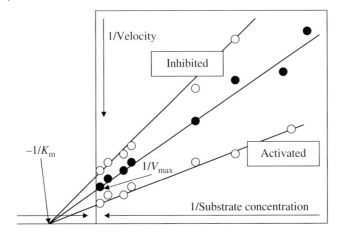

Figure 5.2 Lineweaver–Burk data plot showing a nonmodified enzymatic reaction (dark circles) and two modified enzymatic reactions (clear circles) where one modified reaction is classified as an inhibited enzymatic reaction (V_{max} value is decreased) and the second modified reaction is classified as an activated enzymatic reaction (V_{max} value is increased) relative to the V_{max} of the nonmodified enzymatic reaction.

noncompetitive inhibitor and the noncompetitive activator are changed, being higher up the y axis in the noncompetitive inhibitor situation and lower down on the y axis in the noncompetitive activator situation. Remember, Lineweaver–Burk data plots present the reciprocal values of both V_{max} and K_m, thus a larger V_{max} will provide for a smaller value for the data point (i.e., $1/V_{max}$). As with competitive activators, we can illustrate mechanisms for the enzymatic conversion of substrate to product using equations similar to Equation 5.1. In Equation 5.2 we may utilize a similar set of modified mechanisms to illustrate inhibition and activation by a noncompetitive modifier. However, in Equation 5.2 we will let the modifier bind to the enzyme–substrate (ES) complex rather than to free enzyme (E).

As illustrated, Equation 5.2 is incomplete and we need to further define the relationships of the rate constants. We must first, however, remind the reader that the binding of modifier to the enzyme–substrate complex (ES) is again not reversible, as indicated by a unidirectional arrow in Equation 5.2. In that we have been operating under quasi-equilibrium assumptions for the sake of simplicity, let's continue to operate under quasi-equilibrium assumptions. We can do this by setting specific relationship for those rate constants within each of the three mechanisms illustrated in Equation 5.2. These relationships will be as follows:

k_1 is equal to or greater than k_{-1} and both are much much greater than k_2, which is more easily written as:

$$k_1 \geq k_{-1} \gg k_2$$

Similarly:

$$k_3 \geq k_{-3} \gg k_4 \text{ and } k_5 \geq k_{-5} \gg k_6$$

If we now set k_1 equal to k_3 and k_5 (or $k_1 = k_3 = k_5$) and also k_{-1} equal to k_{-3} and k_{-5} (or $k_{-1} = k_{-3} = k_{-5}$), we are stipulating that the equilibrium binding of substrate to enzyme for all three of these mechanisms (the nonmodified, the activated, and the inhibited enzymatic reactions) are the same and the data plots will all intersect at the same point on the x axis. If as we have agreed, the rate constants k_2, k_4, and k_6 do not contribute significantly to the K_m values under quasi-equilibrium assumptions, we are basically saying that neither modifier is modifying the apparent K_m values for the conversion of substrate to product.

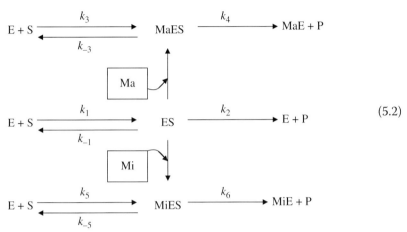

$$(5.2)$$

These assumptions about K_m for the nonmodified and modified mechanisms as illustrated in Equation 5.2, thus make it fairly simple to define noncompetitive inhibitor and noncompetitive activator mechanisms using Equation 5.2 in conjunction with the additional stipulations as defined by the relationships of the rate constants.

Based on the mechanisms illustrated in Equation 5.2, we can simply stipulate that k_4 is greater than k_2, and k_6 is smaller than k_2 (Equation 5.2). This would mean that the MaES complex will break down to the MaE complex and product (P) more rapidly than the ES complex will break down to free enzyme (E) and product (P), and that the MiES complex will break down to the MiE complex and product (P) more slowly than the ES complex will break down to free enzyme (E) and product (P). We thus have inhibition of the enzymatic conversion of substrate to product via one mechanism, and activation of the enzymatic conversion of substrate to product via the second mechanism.

Once again, there are (multiple) other means and mechanisms that can be generated to illustrate noncompetitive modification of an enzymatic reaction. However, the intent here is to get you to think about the roles of the respective rate constants in "defining" K_m and V_{max} values such that you will begin to be able to develop and support your own mechanisms to provide for noncompetitive modification of an enzymatic reaction where the calculated K_m remains the same, but the calculated V_{max} changes in the presence of a modifier. One might quickly realize, for example, that subjecting a solution of enzyme to boiling water might constitute a noncompetitive inhibitor. The high temperatures associated with exposure to boiling water might be expected to inactivate some of the enzyme molecules (thus reducing the V_{max} value); however, those enzyme molecules that were not inactivated will most certainly retain their original K_m values. This simplistic example of a noncompetitive inhibitor (boiling water) could be diagrammed as a reduction in the number (concentration) of free enzyme such that at each available substrate concentration there is a corresponding lesser amount (concentration) of the enzyme–substrate complex to be multiplied by the unchanged rate-limiting rate constant k_2. Thus the overall rate (velocity) of the conversion of substrate to product will be less at each substrate concentration; however, the rate constants k_1 and k_{-1} remain unchanged.

$$k_4 > k_2 > k_6 \tag{5.2.1}$$

Uncompetitive Modifiers

I have saved the most difficult to explain and understand modifiers for last. Uncompetitive modifiers are those solutes that change both the calculated K_m and calculated V_{max} values for an enzymatic reaction. It is possible to illustrate this kind of enzymatic modification in as simple a means as has been used to illustrate competitive and noncompetitive modifiers. We will take this easy route first; however, the reader should be aware that this mode of enzymatic modification affords a teacher the greatest means for challenging the capabilities of a student in that there are so many possible mechanisms that it is virtually impossible to cover all options.

Let's utilize the Lineweaver–Burk data plot to illustrate uncompetitive modification of an enzyme (Figure 5.3). The inhibited and activated data plots in Figure 5.3 are shown as being roughly parallel to the nonmodified enzymatic

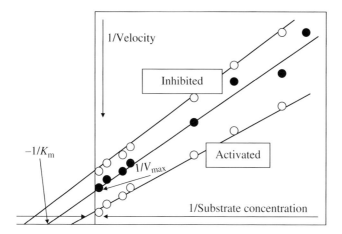

Figure 5.3 Lineweaver–Burk data plot showing a nonmodified enzymatic reaction (dark circles) and two modified enzymatic reactions (clear circles) where one modified reaction is classified as an inhibited enzymatic reaction (V_{max} value is decreased and K_m value is decreased) and the second modified reaction is classified as an activated enzymatic reaction (V_{max} value is increased and the K_m value is increased) relative to the V_{max} and K_m values of the nonmodified enzymatic reaction.

reaction at each substrate concentration. For purposes of selecting a simple mechanistic explanation, I have illustrated the two modified lines as roughly parallel (not crossing other lines), which is about as simple as we can get and still enable us to follow the previous strategy of examining changes in rate constants in additional mechanisms, whereby a modifier changes the respective rate constants such that both K_m and V_{max} values are changed.

By now you should be familiar with the strategy for illustrating modified enzymatic reactions by using an inhibited and an activated mechanism that is derived from a nonmodified mechanism such as illustrated in Equations 5.1 and 5.2. So, let's create Equation 5.3 by simply copying Equation 5.1:

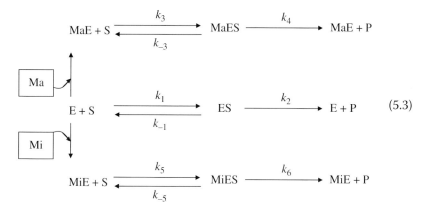

$$(5.3)$$

As you can see, Equation 5.3 is the same as Equation 5.1. However, you will by now also quickly realize that Equation 5.3 is incomplete without defining the relationships of the rate constants.

Therefore, let's establish quasi-equilibrium assumptions once again by stipulating that as illustrated, Equation 5.3 is incomplete and we need to further define the relationships of the rate constant. We must first, however, remind ourselves that the binding of modifier to free enzyme is not reversible, as indicated by a unidirectional arrow in Equation 5.3. In that we have been operating under quasi-equilibrium assumptions for the sake of simplicity, let's continue to operate under quasi-equilibrium assumptions. We can do this by setting specific relationship for those rate constants within each of the three mechanisms illustrated in Equation 5.3. These relationships will be as follows:

k_1 is equal to or greater than k_{-1} and both are much much greater than k_2, which is more easily written as:

$$k_1 \geq k_{-1} \gg k_2$$

Similarly:

$$k_3 \geq k_{-3} \gg k_4 \text{ and } k_5 \geq k_{-5} \gg k_6$$

So, how can we make the mechanisms illustrated in Equation 5.3 meet the needs of an uncompetitive modifier? The requirements of an uncompetitive modifier are that both the calculated K_m and calculated V_{max} in a modified enzymatic reaction be changed relative to the nonmodified mechanism. We can simplify our answer by first stating that k_{-1} equals k_{-3} and k_{-5}. This means that the breakdown of the nonmodified ES complex back to free enzyme (E) and substrate (S) occurs at the same rate as the breakdown of the modified MaES and MiES complexes to their respective modified enzymes (MaE and MiE) and substrate (S). Once we have stipulated these relationships (once we have taken an "easy" option), all we have to state is that k_1 does not equal k_3 or k_5 AND that k_2 does not equal k_4 or k_6. Once we stipulate this additional set of relationships of rate constants, we have immediately stipulated that the modified enzymatic reactions (whether an inhibition or activation modification) will provide for calculated K_m and calculated V_{max} values that are different from the calculated K_m and calculated V_{max} value of the nonmodified enzymatic reaction.

The devil, as they say, is in the details and now it is incumbent on you to figure out whether each respective rate constant in the modified enzymatic reactions needs to be larger or smaller than its corresponding rate constant in the modified reaction to render the modification an inhibition or an activation. In some circumstances, it may also be incumbent on you to determine what you mean by "inhibition" and "activation", but for now let's make something up to illustrate points.

Activation

Let's stay with the information in Figure 5.3 and Equation 5.3 for now and think about activation of the enzymatic reaction. As you can see in Figure 5.3, the activated enzymatic reaction exhibits a V_{max} value that is larger than the corresponding V_{max} in the nonmodified enzymatic reaction. The K_m value for the activated enzymatic reaction is also larger than the K_m in the nonmodified enzymatic reaction; for example, the K_m of the nonmodified reaction can be said to be 1×10^{-6} and the K_m of the activated reaction can be said to be 1×10^{-3}.

How to figure out the relationships of k_4 relative to k_2? Is k_4 larger or smaller than k_2? In that we are under quasi-equilibrium assumptions we have already agreed that the velocity is equal to the rate-limiting rate constant in the simple mechanism times the concentration of the enzyme substrate complex (whether the enzyme is modified or not):

$$\text{Velocity } (v) = k_2 \left([\text{ES}] \right)$$

and

$$\text{Velocity } (v) = k_4 \left([\text{MaES}] \right)$$

At maximal substrate concentration, velocity will equal V_{max} and thus all we have to stipulate to get a V_{max} from $k_2([\text{ES}])$ to be larger than the V_{max} from $k_4([\text{MaES}])$ (because at infinitely high concentrations of substrate the concentration of ES will be equal to the concentration of MaES), is state that k_4 is larger than k_2.

Thus, we have a modified (activated) enzymatic reaction where the calculated K_m of the reaction is larger than the calculated K_m of the nonmodified reaction, and a calculated V_{max} of the reaction that is larger than the V_{max} of the nonmodified reaction. Moreover, we have met the specified conditions for an uncompetitive activation by an activator molecule. We have a modified enzymatic reaction that requires a greater concentration of substrate to achieve a

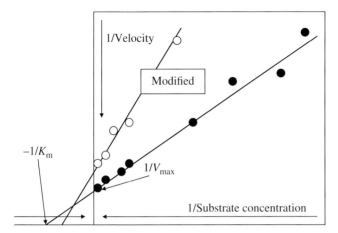

Figure 5.4 A Lineweaver–Burk data plot showing a nonmodified enzymatic reaction (dark circles) and one modified enzymatic reaction (clear circles) where the modified reaction is classified as either an inhibited enzymatic reaction (K_m value is increased relative to the nonmodified reaction) or as an activated enzymatic reaction (V_{max} value is increased) relative to the V_{max} of the nonmodified enzymatic reaction.

measured velocity for an enzymatic reaction at each concentration of substrate tested. We also have a modified enzymatic reaction that will ultimately achieve a higher maximal rate of conversion of substrate to product at infinitely high concentrations of substrate. If you will look back to Figures 4.5–4.7 you can see those enzymatic reactions in the aromatic amino acid biosynthetic pathway that are activated, that is, labeled as "(pos)" in the figure. I am not implying here that these enzymatic reactions are being activated by uncompetitive modifiers. I only seek to link the information in Figure 5.3 and Equation 5.3 to a previously illustrated metabolic pathway.

Inhibition

Let's turn our attention now to the uncompetitive inhibitor illustrated in Figure 5.4. As you can see, the inhibited enzymatic reactions associated with the data plots in Figure 5.4 form a straight line on the Lineweaver–Burk plot where the line intersects the vertical (*y*) axis above where the nonmodified data plot intersects the vertical (*y*) axis AND it intersects the horizontal (*x*) axis to the left of where the nonmodified data plot intersects the horizontal (*x*) axis. The points of intersect of the inhibited enzymatic reaction indicate a lower V_{max} and a smaller K_m for the inhibited enzymatic reaction as compared to the nonmodified enzymatic reaction.

If you grasped the information provided for an uncompetitive activator, you already know that for an uncompetitive inhibitor in Equation 5.3, k_6 will be smaller than k_2. Remember, under quasi-equilibrium assumptions we have the following equations:

$$\text{Velocity } (v) = k_2([ES])$$

and

$$\text{Velocity } (v) = k_6 \left([\text{MiES}]\right)$$

At infinitely high substrate concentrations, the V_{max} of the inhibited reaction will be less than the corresponding V_{max} of the nonmodified enzymatic reaction since the concentrations of ES and MiES will be the same because k_6 is less than k_2.

Determination of the relationship of k_1 and k_5 is similarly easy to figure out from Figure 5.3. We can see that the line for the data plot of the inhibited enzymatic reaction intersects the horizontal line to the left of the point of intersection of the nonmodified enzymatic reaction on the x axis. This means that the K_m of the inhibited enzymatic reaction is smaller than the K_m of the nonmodified enzymatic reaction; for example, the K_m of the nonmodified reaction changes from 1×10^{-6} to 1×10^{-9}. In this example of K_m changes, it now takes a 1000-fold lower substrate concentration to achieve a calculated velocity (rate) of the modified enzymatic reaction equal to half the V_{max} relative to the substrate concentration required to achieve a calculated velocity (rate) of the nonmodified enzymatic reaction equal to half the V_{max}, but since the V_{max} will be less, we still have an "inhibited" mechanism.

Remember, that under quasi-equilibrium assumptions, the following relationships with respect to K_m values exist:

$$K_m = \frac{k_1}{k_{-1}}$$

and

$$K_{mi} = \frac{k_5}{k_{-5}}$$

By *making* k_{-1} equal to k_{-5}, changes to k_1 and/or k_5 directly impact on the K_m values of the respective enzymatic mechanisms, and the rate constants k_2 and k_6 have only minimal impact on K_m because these latter two rate-limiting rate constants are extremely small compared to the equilibrium rate constants.

Another Uncompetitive Modifier

While we are on the topic of uncompetitive modifiers, let's think about another possibility (option). Let's consider the situation illustrated in Figure 5.4.

If you recall an earlier point about the devil being in the detail, you are probably thinking that the devil has also entered into the writing of this book. There is no question about the modified enzymatic reaction being modified by an uncompetitive modifier. The enzymatic reaction that has been modified clearly presents both K_m and V_{max} values that are different from the K_m and V_{max} values of the nonmodified enzymatic reaction. The question, however, is whether the uncompetitive modifier is an activator or an inhibitor?

In this particular instance, the K_m value of the modified enzymatic reaction is larger (intersects to the right of the point of intersection of the nonmodified enzymatic reaction) than the K_m of the nonmodified enzymatic reaction, and

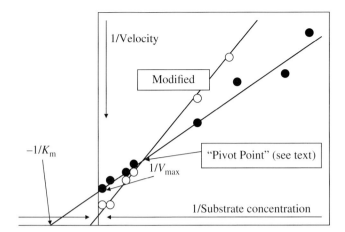

Figure 5.5 A Lineweaver–Burk data plot showing a nonmodified enzymatic reaction (dark circles) and one modified enzymatic reaction (clear circles) where some modified reactions would be classified as an inhibited enzymatic reaction (measured velocities are reduced relative to nonmodified enzyme velocities), but other modified reactions would be classified as an activated enzymatic reaction (measured velocities are increased relative to nonmodified enzyme velocities). Both calculated V_{max} and K_m values of the modified enzymatic reaction are changed relative to the V_{max} and K_m values of the nonmodified enzymatic reaction, that is, an uncompetitive modification mechanism.

thus it now takes greater substrate concentrations to achieve a measured velocity equal to half the V_{max} for the modified enzymatic reaction than for the nonmodified enzymatic reaction (an "inhibitor"?). The modifier has decreased the ability of the substrate to "bind to" the substrate-binding site and thus is behaving similarly to the inhibitor in Equation 5.3 (Figure 5.3). However, the modifier is also reducing the measured velocities of the enzymatic reaction at each substrate concentration tested (analogous to the inhibitor in Equation 5.1 and Figure 5.1) and we would consider this to be an inhibition of the enzymatic reaction and the modifier to be an inhibitor. In fact, this situation is relatively simple in that the modifier reduces (a greater concentration of substrate) the value of the K_m and reduces the value of the V_{max} and thus is clearly an uncompetitive inhibitor. Can you determine the relationships of the rate constants k_1, k_{-1}, k_2, k_5, $k_{-;5}$, and k_6 from Equation 5.1 that will satisfy the needs of this modified (inhibited) enzymatic reaction?

Ok, let's look at another possibility (option). Let's look at Figure 5.5. The modified enzymatic reaction data plot in Figure 5.5 shows the intersection of the straight line as being lower on the vertical (y) axis (greater V_{max}) than the point of intersection of the straight-line data plot derived from the nonmodified enzymatic reactions. The data plot for the modified enzymatic reaction also intersects the horizontal (x) axis to the right (larger K_m value, i.e., greater substrate concentration) of the point of intersection of the data plot for the nonmodified enzymatic reaction. We appear to have a modifier that decreases the (provides for a larger) calculated K_m for the modified enzymatic reaction, but also increases the calculated V_{max} value of the modified enzymatic reaction relative to the V_{max} value of the nonmodified enzymatic reaction. We also note that the individual data points for the modified enzymatic reaction fall both above and below comparable (at the same or similar substrate concentrations) data points for the nonmodified enzymatic reaction. At low substrate concentrations, the modifier appears to inhibit the enzymatic reaction, but at higher substrate concentrations the modifier appears to activate the enzymatic reaction. How can this be? Can you devise an equation to illustrate a series of mechanisms, all tied to some nonmodified enzymatic mechanism, under quasi-equilibrium assumptions, that will explain how this particular set of data points might be obtained?

Let me start you out with one possible example (Equation 5.4). Equation 5.4 has been reduced to only two mechanisms. One mechanism represents the non-modified enzymatic reaction and the second mechanism represents the modified enzymatic reaction. The equation indicates that substrate (S) reversibly binds to some substrate-binding site on free enzyme to form an ES complex, which is ultimately converted to product. The equation also indicates that substrate (S) can reversibly bind to a second site on free enzyme (E) to form an SE complex (note the difference between SE and ES here) and that this SE form of the enzyme can then reversibly bind substrate (S) to the substrate-binding site forming an SES complex such that substrate is ultimately converted to product via this second mechanism. The equation also indicates that some modifier (M) can irreversibly bind to the SES complex to form the MSES complex, which breaks down to MSE and product. The equation also indicates that the SE complex is re-formed from the MSE complex after the formation of product (P).

Too complicated you say? I agree, but remember that the objective here is to get you to think about how you can manipulate mechanisms and rate constants to obtain some desired graphical plot of velocity versus substrate concentration data (which are the kinds of data you will be seeing in real life). If you try always to explain data plots utilizing the traditional explanations for enzymatic reactions, you will always be lost somewhere between getting a right answer and not getting a right answer. You will be on that part of the map labeled "Beyond this point there be dragons."

Ok, let's return to Equation 5.4 and Figure 5.5. How do the combinations of rate constants and possible mechanisms for the conversion of substrate to product "work"? You will note that there were no rate constants applied to the binding of substrate (S) to free enzyme (E) to form the substrate–enzyme (SE) complex or to the breakdown of this SE complex back to free enzyme (E) and substrate (S). In the binding of substrate to enzyme at a site other than the substrate-binding site, the binding of substrate to this second (modifier?) site is an equilibrium condition and substrate is not converted to product when bound at the second (modifier?) site. What this means is that substrate concentration will serve only to shift free enzyme back and forth between two forms of "free enzyme," free enzyme E and free enzyme SE. Thus the concentration of substrate in the system will impact on which of the two forms of free enzyme will bind substrate, and thus which enzymatic mechanism will actually lead to the conversion of substrate to product. If we set the K_{eq} of the binding of substrate (S) to free enzyme (E) to form the SE complex to a very small value, the pivot point (if I may call it a pivot point—see Figure 5.5) at which the straight line associated with the modified enzymatic reaction crosses the straight line for the nonmodified enzymatic reaction in Figure 5.5 shifts to the right. If we set the K_{eq} to be larger, the pivot point will shift to the left.

By obliging the modifier (M) to irreversibly bind to the SES complex and requiring that the MSES complex irreversibly breaks down to product (P) and MSE—the latter of which is then free to dissociate (break down) to free enzyme (SE) and modifier M, permitting the modifier to recycle back and forth between being bound to some substrate–enzyme–substrate (SES) or enzyme–substrate complex (SE)—we have a mechanism we can deal with.

So, can we think our way through this proposed mechanism(s)? At low concentrations of substrate, substrate will be converted to product primarily through the nonmodified enzymatic mechanism. Thus, at low substrate concentrations the presence of the modifier (M) agent will have little impact on the system as there will be very little substrate–enzyme–substrate (SES) complex for it to bind to and the conversion of substrate to product will not appear to be modified at all. However, as we increase the concentration of substrate, the modified enzymatic mechanism will become more dominant, and at infinitely high concentrations of substrate, virtually all substrate will be converted to product via the modified enzymatic reaction. How do we work with our rate constants to insure that at the higher substrate concentrations the enzymatic conversion of substrate to product appears to be activated, and at low substrate concentrations the enzymatic conversion of substrate to product appears to be inhibited? If we set k_4 to be larger than k_2 we will oblige a faster rate of conversion of substrate to product at the higher substrate concentrations, providing for a larger V_{max} value in the presence of modifier (remember, SES does not convert to product, only to SE and S). How do we make the calculated K_m smaller in the presence of modifier? If we set k_3 to be smaller than k_1 (and make k_{-3} equal to k_{-1}), we are insuring that the binding of substrate to the substrate–enzyme complex will occur more slowly than the binding of substrate to the free enzyme, and since both ES and SES break down equally fast, it will require higher concentrations of substrate (S) to achieve $\frac{1}{2}V_{max}$ for the conversion of substrate to product.

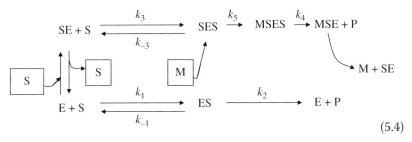

$$(5.4)$$

Are you having fun yet? If you think that there is, or are, easier ways to explain the data plot in Figure 5.5 (or Equation 5.4) or think my proposed mechanisms and rate constant relationships are wrong, then I will have to suggest to you that I have achieved my objective. To date, we do not know a single mechanism by which an enzyme converts substrate to product. In addition, you will encounter (assuming you go on to study enzymatic reactions) information and strange data plots from your studies that will not conform to what you will be taught in enzymology classes as possible mechanisms. It is important that you learn now to think about how you might explain the data you will collect and avoid attempting to fit a square peg into a round hole. If after reading this section of my book, you can think of more than one way to explain the velocity versus substrate concentration data plots, both in the presence and/or absence of things that you think might be modifying the enzymatic reaction, and you can apply different mechanisms to that thinking, you will be ahead of many people who are teaching biochemistry classes.

So, how does all of this modification of enzymatic reactions relate to the water distribution system in Chapter 4 with its vertical and horizontal pipes for the distribution of that water? How can we tie the apparent modifications of enzymes' K_m and V_{max} values to the changes in flow of substrates (nutrient molecules) to different metabolic pathways and thus eventually to the formation of different products produced by living cells?

6

Modification of Metabolite Flow Through Metabolic Pathways

I like nonsense, it wakes up the brain cells. Fantasy is a necessary ingredient in living, it's a way of looking at life through the wrong end of a telescope. Which is what I do, and that enables you to laugh at life's realities.

Dr Seuss

Let's return for a moment to Figure 4.1 (reproduced here as Figure 6.1) so that we may begin to consider how we can change the horizontal pipe configuration to correspond to the changes in K_m and V_{max} we have been discussing in the previous chapter. Let's make a few minor changes to Figure 6.1, presenting these changes in a new figure (Figure 6.2).

If we compare the height and inside diameter of horizontal pipe B in Figure 6.1 with that same horizontal pipe in Figure 6.2 we note two things: pipe B in Figure 6.2 is smaller in diameter (meaning a lower water carrying capacity—analogous to a smaller V_{max} potential as an enzyme) and it is higher up on the vertical pipe (meaning a higher level of water in the vertical pipe is needed before water will even start flowing through the horizontal pipe B—analogous to a larger K_m value for an enzyme). In practical terms, this means that the household (presumably of a royal family if you remember from earlier) serviced by horizontal pipe B is going to receive water second to the household being served by pipe A, *and* it means that household B will also receive less total water than household A over some given time interval (the fountains in household B will not spurt as high as the fountains in household A). In ancient Rome, this would presumably have required two events: 1) the royal family falling into disrepute with the emperor, and 2) a bunch of slaves having to dig up the old horizontal water line B and replacing it with a smaller diameter water line that was higher (or later in the scheme of water flow) on the water distribution system. In present-day Rome, these same two events might correspond to: 1) the water department not receiving the monthly payment of the water bill, and 2) the water department personnel coming out and installing a smaller water meter with a pressure regulator that required a higher water pressure to open and deliver water to the house. In short, the water department personnel correspond to the modifier in an enzymatic reaction and the modifier can modify the enzyme to "deliver more or less water at some greater or lesser water pressure" to the metabolic pathway being serviced.

Enzyme Regulation in Metabolic Pathways, First Edition. Lloyd Wolfinbarger.
© 2017 John Wiley & Sons, Inc. Published 2017 by John Wiley & Sons, Inc.

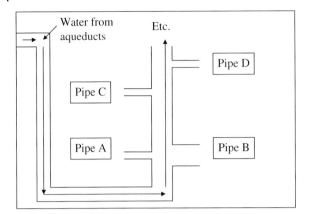

Figure 6.1 A diagrammatic illustration, also shown in Figure 4.1, of a hypothetical water distribution system such as may have been in operation in ancient Rome. The water in this hypothetical system is delivered via an aqueduct from some water source high in the mountains surrounding Rome and enters a water distribution system that feeds water to homes and public facilities in Rome. Pipes A through D are intended to represent volumes of water that might be delivered based on two aspects of the distribution system: 1) the diameter, and thus carrying capacity, of the pipe feeding the home or public facility; and 2) the height of the pipes relative to other pipes in the water distribution system.

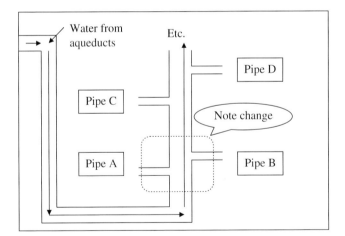

Figure 6.2 A second diagrammatic illustration of a hypothetical water distribution system such as may have been in operation in ancient Rome. However this water system has been changed to reflect changes in the horizontal pipes compared with those same pipes in Figure 6.1. The water in this hypothetical system is delivered via an aqueduct from some water source high in the mountains surrounding Rome and enters a water distribution system that feeds water to homes and public facilities in Rome. Pipes A through D are intended to represent volumes of water that might be delivered based on two aspects of the distribution system: 1) the diameter, and thus carrying capacity, of the pipe feeding the home or public facility; and 2) the height of the pipes relative to other pipes in the water distribution system.

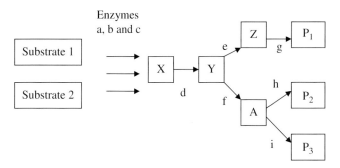

Figure 6.3 Diagram illustrating a metabolic pathway whereby two substrate molecules are metabolized to compound X via enzymes a, b, and c. Compound X is metabolized to compound Y by enzyme d. Compound Y is metabolized to compounds Z and A by enzymes e and f, respectively. Compound Z is metabolized to product P_1 by enzyme g and compound A is metabolized to products P_2 and P_3 by enzymes h and i, respectively.

How does this analogy relate to the flow of metabolites through intermediary metabolism in a living cell? Let's again refer to a previously utilized figure (Figure 4.3, repeated here as Figure 6.3) so that you do not have to search back through the book to review it.

We can use the aromatic amino acid biosynthetic pathway to illustrate one underlying principle in the control of flow of metabolites through intermediary metabolism because this particular pathway, in addition to leading to the synthesis of aromatic amino acids and associated intermediates, also leads to the synthesis of individual amino acids (tyrosine, tryptophan, and phenylalanine) and thus we have several individual pathways that come off of one central pathway. Thus, for purposes of our water distribution system analogy, the central pathway represents the vertical water pipe and the several individual pathways represent the horizontal water pipes in our earlier representations of a water distribution system in ancient Rome.

We see from Figure 6.3 that there is what we will call a "commitment step" in the aromatic amino acid biosynthetic pathway. This enzyme is D-arabinoheptulo-sonate-7-phosphate synthetase—the enzyme converting erythrose-4-phosphate (E4P) and phosphoenolpyruvate (PEP) into D-arabinoheptulosonate-7-phosphate (DAHSP)—and there are three isozymes of this enzyme. Each isozyme catalyzes the same conversion of substrate(s) to product, but each isozyme is different from the others by virtue of which modifiers affect the enzymatic reaction and which kind of modification they affect (look back to Figures 4.5–4.7 to see how each aromatic amino acid modifies a given isozyme). If an abundance of the amino acid phenylalanine were to become available to the cell, there would be little need for the cell to synthesize phenylalanine. In this pathway, phenylalanine feedback (meaning an end product impacts some enzymatic reaction that occurs along its biosynthetic pathway) inhibits (modifies) one of the isozymes and has the potential to activate (modify) two other of the isozymes. Insofar as this binding of an end product of the metabolic pathway is not at some substrate-binding site, the binding of modifier to enzyme is presumably at a modifier site and thus the binding component of the event involves only an association constant and a dissociation constant, that is, a K_{eq} constant. The ratios of these association and dissociation constants define the concentration of the end product (in this case phenylalanine) that will be required to minimally, partially, or completely occupy the modifier-binding sites of all such isozymes capable of binding this particular modifier. If high concentrations of phenylalanine are present, presumably virtually all of the isozymes capable of binding the phenylalanine will be modified and the enzymatic conversion of E4P and PEP to DAHSP will be modified (inhibited?).

The question, however, is how will the enzymatic conversion be inhibited (modified)? Will the modifier (phenylalanine) act by changing the calculated K_m (increasing the concentrations of E4P and PEP required to achieve $\frac{1}{2}V_{max}$) (i.e., raising the pipe higher in the water distribution system), or by decreasing a rate-limiting rate constant such as k_2 in our examples (using a smaller diameter pipe in the water distribution system)?

Having inhibited one of the three isozymes in the commitment step of aromatic amino acid biosynthesis, we might find that that we are left with two "minor" issues. Firstly, if the metabolic activity is producing erythrose-4-phosphate and phosphoenolpyruvate at some steady state and we remove an enzymatic reaction that is consuming some of the two molecules being produced, we might expect some greater amounts (concentrations) of these two metabolites to be accumulating. Secondly, we might expect a lesser amount (lower concentration) of D-arabinoheptulosonate-7-phosphate (DAHSP) to be produced, and this lesser amount (lower concentration) is going to impact the enzymatic conversion of this particular metabolite to shikimic acid (actually a metabolite intermediate between the DAHSP and shikimate), which is going to impact the enzymatic conversion of shikimate to chorismate, and so on and so forth. Just because the cell has an abundance of phenylalanine does not mean it does not need to continue the synthesis of tyrosine and tryptophan. Thus, we have the ability of phenylalanine to modulate the other two isozymes as an activator such that the whole system is capable of modulating all three isozymes, maintaining some new steady-state conversion of E4P and PEP to DAHSP.

You will note that phenylalanine also feedback inhibits the enzyme converting prephenic acid to phenylpyruvate (and ultimately to phenylalanine—enzyme h in Figure 4.6) and activates the conversion of prephenic acid to hydroxyphenylpyruvate (and ultimately to tyrosine—enzyme i in Figure 4.6). Phenylalanine also activates enzyme e, which is involved in the conversion of chorismic acid to anthranylic acid (anthranilate), which will ultimately be converted to the amino acid tryptophan.

I could repeat this same description of regulation of the synthesis of the aromatic amino acids by excess amounts of the end products tyrosine and tryptophan, but I think you can track these events yourself by looking carefully at Figures 4.5–4.7. The take-home lesson is that in this particular metabolic pathway, the end products of this pathway regulate the flow of metabolites within and through the pathway by selectively modifying (either activation or inhibition) specific enzymes within that pathway. Where there is an abundance of one particular aromatic amino acid, that aromatic amino acid is able to reduce the levels of the initial intermediates in the pathway and direct what intermediates are still made toward the other aromatic amino acids. This would be analogous to the changes in the diameters and/or heights in water pipes distributing water to homes based on consumption, where if one home didn't need the water, water could be redirected to other homes. Obviously cells can modulate the levels of enzymes within metabolic pathways by controlling the levels of transcription of specific genes and/or translation of the mRNAs that code for specific enzymes; however, the objective of this book is to deal with regulation of the enzymes themselves by modifiers.

Just as you thought I was finishing with this chapter, I have taken Equation 5.3 from Chapter 5 and inserted it here as Equation 6.1. However, in this part of our discussion on regulation of a given enzyme in the aromatic amino acid biosynthetic pathway, I would like to take you back down memory lane and ask that you think about: 1) how an abundance of the aromatic amino acid phenylalanine might modify ("inhibit") an isozyme (i.e., enzyme "b") the enzyme D-arabinoheptulosonate-7-phosphate synthetase (DAHPS—the enzyme that combines erythrose-4-phosphate (E-4-P) and phosphoenolpyruvate (PEP), i.e., the two substrates in Figure 5.3); and 2) how an abundance of the aromatic amino acid tryptophan can also modify ("activate") this same isozyme (i.e., enzyme "b"). You will need to look back specifically at Figures 4.6 and 4.7—specifically at the dashed lines pointing to the isozyme (middle arrow) between the substrates E-4-P and PEP and the "product" shikimate (which would correspond to enzyme "b"). In this case, regulation of this particular isozyme can be both inhibited by phenylalanine and activated by tryptophan (I won't confuse you further by also saying that tyrosine can also activate this particular isozyme).

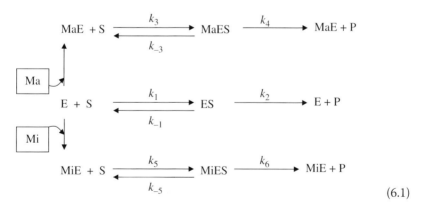

$$(6.1)$$

If you look at Equation 6.1 you will see that it might (just as one example) represent such a regulatory event with enzyme "b" in the aromatic amino acid biosynthetic pathway. In this equation, the modifier "Mi" could be phenylalanine and the modifier "Ma" could be tryptophan? If you accept this option, then in the total absence of aromatic amino acids for regulation of the aromatic amino acid biosynthetic pathway, the unmodified mechanism should function for the DAHPS enzyme. If phenylalanine should become available in abundance, we would expect that the DAHPS enzyme would shift to the bottom pathway (the "Mi" modified pathway), but should an abundance of tryptophan become available we would expect that the DAHPS enzyme would shift to the top pathway (the "Ma" modified pathway). The question for you, however, is since I have also described for you that tyrosine will also be activating this particular isozyme (see Figure 4.5), what kind of mechanism would you envision should an abundance of both tryptophan and tyrosine occur (without an abundance of phenylalanine)? Would you think a second modifier (Ma2 versus perhaps just Ma) would bind to free enzyme along with the first modifier to make a "Ma–Ma2E" complex? Would you think that a second modifier might bind to free E such that it precluded the first modifier from binding so that collectively you would get two kinds of modified enzyme complex, such as "MaE" *and* "Ma2E," each with differing sets of rate constants for the conversion of substrate to product? Can you work your way through the other two isozyme forms of the DAHPS enzyme (i.e., enzymes "a" and "c")? Would you expect similar/different strategies for the modification of these two isozymes? It should be obvious to you by now that you should consider each enzymatic step in the regulation of the aromatic amino acid biosynthetic pathway (as well as all other enzymatic pathways and/or individual enzymes) as being modified and be prepared to anticipate how the regulation (modifications) might be accomplished and how you can best prepare your experimental designs when approaching a study of some enzyme-driven chemical reaction. How would you adjust the respective rate constants in a given equation to manipulate the apparent K_m and/or V_{max} values of some modified enzymatic reaction? Can such a modified reaction be "activated" by decreasing the apparent K_m or by increasing the apparent V_{max}, and how would you need to change the respective rate constants to achieve such a modification (regulatory) event? I promise you, it will never be easy and the kinetics of some enzymatic reaction will always be more complicated than you imagined.

It is perhaps important to emphasize at this point that regulation of the flow of metabolites within a living cell is extremely complex and I am speaking only to a very small part of that regulation. The water distribution example I have chosen to use is a good analogy in that you can think of the vertical pipe in the examples given as representing a total pool of a given metabolite. Each metabolite in intermediary metabolism has the potential to be used in several enzymatic pathways to generate multiple other metabolites (we might refer to these as end products if it helps you to think about the overall systems). Whether the starting intermediate flows to one pathway in preference to another pathway is mostly modulated by the K_m of that "committing" enzyme of that metabolic pathway, and the amounts of metabolite that flow into and through that pathway are modulated by the V_{max} of that committing enzyme. It is therefore easy to say (with some degree of certainty) that the intracellular concentration of a given metabolite will almost always approximate the average K_m values of those enzymes drawing on that metabolite as a substrate. Having a given metabolite at or near the K_m values of enzymes metabolizing that metabolite provides for the most "sensitive" regulatory event, that is, where small changes in concentration(s) will have the greatest effect on enzymatic conversion(s) (refer back to the initial slope on a velocity versus substrate concentration data plot in Figure 3.1). In addition, we should expect that regulation of a given enzyme will, most likely, modulate any changes in the K_m of an enzyme to one order of magnitude above and below the value of that K_m. Modulating a K_m change greater than this would be of minimal value because a change of one order of magnitude in some K_m value will approximate a 90% change in measured velocity of that reaction in any effect that changing a K_m will achieve.

7

Which is the Real Substrate?

Where is the wisdom we've lost in information?

T.S. Elliot

As well as looking at how one determines the rates of conversion of substrate to product in order to establish how K_m and V_{max} values manage intermediary metabolism, as described in Chapter 6, one has to pay attention to knowing what the real substrate might be in an enzymatic reaction. For example, consider the lowly amino acid, aspartic acid. In an aqueous environment, aspartic acid exists in four different molecular forms (species). There is the form that carries a double negative charge (at high pH), a form that carries a single net negative charge (at lower pH), a form that is zwitterionic (no net charge), and a form that is positively charged at very low pH. Aspartic acid has three ionizable groups, two carboxyl groups and one amino group. Each ionizable group presents a specific pK value (a value at which a specific ionizable group is half associated/dissociated as a function of the hydrogen ion concentration (pH) in the solution. Again, it is necessary to remind you that ionization of a given ionizable group is never 100% and therefore at every pH value all four forms (species) of aspartic acid will be present in the solution even though the form (species) of any particular charge may be virtually nonexistent (with respect to its concentration) nor will any form (species) ever equal the total molar concentration of aspartic acid dissolved in the aqueous solution.

As may be seen in Figure 7.1, each form (species) of aspartic acid will present differing ionic charge distributions, and since any enzyme that might bind aspartic acid as a substrate might be expected to differentiate between a given ionic species, we might again ask the question: Which form of aspartic acid is the substrate of that enzyme, and what is the actual concentration of that substrate species relative to the total concentration of aspartic acid you have dissolved in the aqueous solution?

Let's try first to deal with the changing concentrations of each molecular species of aspartic acid as a function of pH. It should seem obvious that at extremely low pH, all ionizable groups on the aspartic acid would be protonated. Each ionizable group will present a pK value (which will vary a bit according to the ionic environment and ion composition of the aqueous solution, but which we will ignore for now) and at sufficiently high concentrations of hydrogen ions ($[H^+]$)

Enzyme Regulation in Metabolic Pathways, First Edition. Lloyd Wolfinbarger.
© 2017 John Wiley & Sons, Inc. Published 2017 by John Wiley & Sons, Inc.

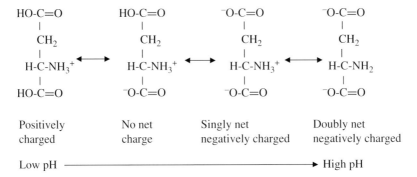

Figure 7.1 Graphic illustrating the four molecular species of aspartic acid as a function of changing pH and ionization of the three ionizable groups.

the equilibrium of the association/dissociation reaction will be shifted towards "association." As the hydrogen ion concentration progressively declines (pH rises), each ionizable group will begin to shift towards dissociation as a function of the pK value of each ionizable group, and eventually at very low concentrations of hydrogen ions (high pH) all ionizable groups will tend toward full dissociation. Thus the concentrations of each molecular species will change as a function of pH. Remember, however, that the total concentration of all four species must always add up to the total concentration of aspartic acid dissolved in the aqueous solution. It is this relationship that permits calculation of the concentrations of each of the four molecular species at any given pH and thus the actual (or real) concentration of the molecular species that represents the true substrate for the enzyme at any given time in an enzymatic reaction.

Figure 7.2 shows the approximate concentrations of each of the four molecular species of aspartic acid. Note that it is only for illustrative purposes since it is "hand drawn" and not intended to show actual concentrations. However, it serves to illustrate the "problem" in dealing with a substrate that can exist in multiple forms depending on a variable such as pH. As you can see in the graphic, depending on the pH at which your enzymatic reaction is being measured, the concentration of the actual substrate species will be different from the concentration of aspartic acid you dissolved into the aqueous solution. Obviously, not knowing the actual (or effective) concentration of the true substrate of an enzymatic reaction will impact the accurate determination of K_m and V_{max} values for that enzyme reaction.

There are multiple teaching points that we can discuss here. It should be obvious to you by now that if you run an enzymatic reaction at a given pH and attempt to measure the conversion of aspartic acid to another compound (perhaps you wish to deaminate aspartic acid to gain a four-carbon dicarboxylic acid—succinic acid) you will have to contend with the presence of four different species of aspartic acid in the reaction. Pretending that you know which molecular species is the true substrate of the enzymatic reaction it should become obvious to you that all four species are in equilibrium with each other, and as you consume substrate (convert to product), the true molecular species that is the substrate of the enzyme will be replenished as molecular species that are not the true substrate

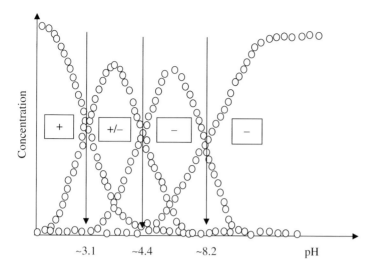

Figure 7.2 Graphic to illustrate how the concentrations of each of the four molecular forms (species) of aspartic acid change as a function of pH.

will redistribute themselves in order to maintain equilibrium. Theoretically, as you remove molecules of one species, the total overall concentration of aspartic acid will decline even though the other three species are not the actual substrate of the enzymatic reaction, but the effective concentration of the substrate that you use to calculate a K_m value may not be sufficient to achieve a maximal rate of conversion of substrate to product. In addition, unless you have a good strong buffer in the system, it is quite possible that the pH of the system may change because the different molecular species present differing pK values for the three ionizable groups on aspartic acid, and thus buffering capacities may change over the course of your assay, further changing the concentrations of each species.

As may be seen in Figure 7.2, which shows how the concentrations of each molecular species of aspartic acid vary as a function of pH, it would be an easy experiment to determine some optimal pH for the enzymatic conversion of aspartic acid to, for example, succinate. "Easy" of course is a matter of whether or not one would have to contend with changes in pH resulting in protonation/deprotonation of ionizable groups on the enzyme catalyzing the reaction. If one did not have to contend with ionizable groups on the enzyme and one used a total concentration of aspartic acid known to be at "saturating" concentrations (e.g., approximately 10 times higher than the determined K_m for the reaction), the optimal pH would obviously be the same as the top of each concentration for each species as illustrated in Figure 7.2. For example, if the zwitterionic species of aspartic acid were the true substrate of the enzymatic reaction, the optimal pH would lie somewhere between pH 3.1 and 4.4 because at any pH above or below the mid-peak concentration for this molecular species, the concentration of this molecular species would decline and the rate of the enzymatic conversion would diminish. Obviously, one cannot disregard ionization of groups on the enzyme, and thus one should expect that changes in pH in measuring the rate(s) of the enzymatic reaction will represent events associated with both the substrate (true substrate?) and the enzyme. While we have focused this discussion on an enzymatic conversion of aspartic acid to some product, it should be obvious that the description of effects of changing pH on enzymatic reactions will apply to virtually any substrate-to-product conversion and especially to those enzymatic reactions involving a substrate or product with at least one ionizable group that involves a protonation/deprotonation event. It just keeps getting better and better.

Finally, I would like to return to my earlier introduction of why nucleic acids are helical, but only to ask a question that enables me to go back to my earlier observation that "The sums of the parts dictate the function(s) of the whole." When I was a student in biochemistry and enzymology classes, I sort of assumed that the polymerase enzyme (the enzyme we are taught polymerizes nucleic acids into double-stranded alpha-helical structures) bound up the nucleotide triphosphates and using a template of the opposite strand began linking those nucleotide triphosphates into a linear polymer. This sort of meant to me that I should have binding sites on the polymerase enzyme for nucleotide phosphates (equal binding for all?) and for the linear single strand of polynucleotides that determined which nucleotide phosphate would best fit the enzyme-binding (and catalytic) site in the polymerase polymerization process.

However, later in life I began to understand that nature seemed to operate on what I have always termed "self-assembly" and that the nucleotide triphosphates would stack like a stack of pennies when in water. Furthermore, given that the bonding of the purines and pyrimidines would dictate the linear sequence of some polymerized nucleic acid, I began to understand that the molecular components (the parts) of nucleic acids possessed the capacity of recognizing the proper way to associate with each other. Thus I began to question whether the polymerase might not need to recognize nucleotide phosphates for binding in some binding site or the opposite strand of a nucleic acid polymer. What if all the polymerase needed to recognize were those aspects of the parts that needed to be covalently linked? What if all the polymerase was doing was providing a hydrophobic region that served a similar function to the nonenzymatic polymerization of nucleic acids as the "parts" self-assembled on the surface of some clay particle and as the water evaporated, chemical bonds were formed? Does the ribosome perform a similar function as it polymerizes amino acids into proteins? What if what we think we know is wrong? Did we ask questions, or did we simply go along with someone else's thoughts because we thought they were smarter?

I'm not trying to challenge or label as untrue the things you have been taught or the things you believe to be true. What I am trying to do is get you to try to think through what you are learning, and question, challenge, and rethink what you think you know and what you are being taught. I have tried to take you through possible enzymatic mechanisms, possible data plots, and how you might apply these possibilities to understand cellular metabolism. I accept that the preceding chapters were challenging and difficult to follow with respect to my thought processes. I hope you challenged and will continue to challenge what I tried to teach you.

8

Non-Quasi-Equilibrium Assumptions

The old adage that you can catch more flies with honey than with vinegar formed the basis for this book. You got the honey, now comes the vinegar.

The Author

I could not leave the discussion of enzyme kinetics without briefly broaching the subject of an enzymatic reaction (equation) that did not satisfy quasi-equilibrium assumptions. I will try to be brief, but it is important that you understand that I used quasi-equilibrium assumptions in this book in order that you would get the honey rather than the vinegar.

The use of steady-state assumptions in deriving rate equations represents a dynamic system where some aspect of a particular quantity remains *constant*. For example, you can think of an analogy where you sell and/or return as many bags of peanuts as get delivered to your store. In this analogy, quasi-equilibrium conditions would exist if you returned most of the bags of peanuts you received and sold only a few bags over that time interval. Steady-state conditions would exist if you sold most of the bags of peanuts you received and returned only a few bags over time. While it is true that the number of bags of peanuts you have in your store remains fairly *constant* in both situations, selling more than you return represents a more dynamic situation for you (you make money). If we repeat Equation 3.2 (as Equation 8.1 here), we can try to explain this difference in a simple change to this equation.

$$S + E \underset{k_{-1}}{\overset{k_1}{\rightleftharpoons}} ES \overset{k_2}{\longrightarrow} P + E \tag{8.1}$$

For steady-state assumptions, all we need to do to modify the characteristics of Equation 3.2/8.1 is to stipulate that k_2 is NOT much much less than k_{-1} and/or k_1. This situation would mimic you receiving over time an amount of "substrate" (the packages of peanuts) pretty much equal to the packages of peanuts (made product) you sold. Fewer bags of peanuts would be returned to the vendor under steady-state assumptions than under quasi-equilibrium assumptions. We can illustrate this situation by relating the relative values of k_1, k_2, and k_{-1} to each other. Obviously we can create a fairly infinite series of relationships where

Enzyme Regulation in Metabolic Pathways, First Edition. Lloyd Wolfinbarger.
© 2017 John Wiley & Sons, Inc. Published 2017 by John Wiley & Sons, Inc.

$k_1 = k_{-1} = k_2$; $k_1 = k_2 >> k_{-1}$; $k_1 < k_2 > k_{-1}$, and so forth (you get the picture). Let's try to think about what these relationships of rate constants mean.

Let's first consider the situation where $k_1 = k_{-1} = k_2$. Let's also first consider that substrate (S) is at "saturating conditions" with respect to enzyme (E) and there is an infinite amount of substrate (it will not be depleted during the course of conversion of substrate to product). Obviously you will be forming the enzyme–substrate complex (ES) at some maximal value, but basically (once the system comes to some steady state—meaning [ES] becomes *constant*) you will be converting as much ES to P as you convert ES to E + S. In this case, k_2 will still be the rate-limiting step in the conversion of substrate to product, but k_{-1} will be a serious contributor to the conversion of substrate to product by limiting the amounts (concentration) of the enzyme substrate complex at equilibrium. In this case, [ES] can never approximate [E_t] like it can under quasi-equilibrium assumptions. It must therefore approximate some aspect of steady-state assumptions. The question is, can you now work your way through some modification of this enzyme mechanism where the nonmodified mechanism in Equation 8.2 can be either activated or inhibited (this is the vinegar)? Remember that $k_1 = k_{-1} = k_2$. Let's start with some modifier acting as an activator. In order to keep this simple, let's assume in Equation 8.2 that $k_3 \geq k_{-3}$ and k_4 is $<< k_3$ (and k_{-3}). This puts the modified mechanism under quasi-equilibrium assumptions, but the nonmodified mechanism remains under steady-state assumptions. Now, what additional parameters do you need to decide on? In Equation 8.2, modifier (Ma) binds irreversibly to free E, so given some excess of modifier (Ma) in relation to free enzyme (E) it should be obvious to you that you will simply give up one velocity versus substrate concentration data plot for a second similar substrate versus substrate concentration data plot (exchange one mechanism for a second mechanism). The differences between the two data plots will, however, differ based on how you relate k_1 to k_3, k_{-1} to k_{-3}, and k_2 to k_4. Ordinarily you might think that if you let $k_1 = k_3$, $k_{-1} = k_{-3}$, and let k_4 be greater than k_2 you would have a simple "competitive activator" (V_{max} changes, but "K_m" remains the same). However, remember that under steady-state assumptions for the nonmodified mechanism, k_2 factors strongly into the K_m, but k_4 factors only minimally in the modified mechanism under steady-state assumptions. You would thus be modifying both the apparent K_m and apparent V_{max} when modifier was added, and although the overall rate of conversion of substrate to product in the presence of modifier will be greater ($k_4 > k_2$), your modifier will now yield a data plot of an uncompetitive modifier. The V_{max} of the modified data plot will be greater (higher up on a V vs S data plot), but will the new K_m value shift to the left or the right on the V versus S data plot? Will the new K_m be representative of a greater or lesser concentration of substrate to achieve a measured velocity equal to one-half of the V_{max}?

Did you figure it out? In the nonmodified mechanism of Equation 8.2, K_m essentially equaled k_{-1}/k_1, but in the modified mechanism in Equation 8.2 K_m essentially equaled $(k_{-3} + k_4)/k_3$. If it helps, let's give actual numbers to the rate constants. Let k_1 and k_3 equal 1000 (remember we let k_1 equal k_3). Let k_{-1} and k_{-3} equal 100 (remember we let k_{-1} equal k_{-3}). Similarly let k_2 equal 10 and k_4 equal 1000 and we will have satisfied the parameters we set in the previous paragraph. So, $100/1000 = 0.1$ (k_{-1}/k_1) and $100 + 1000/1000 = 1.1$ (($k_{-3} + k_4)/k_3$). Is 0.1 M (molar—remember K_m is essentially thought of as a concentration of a substrate that yields a measured velocity equal to one-half V_{max}) larger or smaller than 1.1 M as a "substrate concentration" necessary to achieve a measured velocity equal to one-half the V_{max}? K_m gets "bigger" (modified mechanism requires a greater substrate concentration than the nonmodified mechanism) so the measured "new" K_m will move to the _____. Did you really think I would tell you at this point in the book? What would data plots of velocity versus substrate concentration look like for this modified enzymatic reaction? Would it look like Figure 8.1 without the clear data points (and data line) below the black data points on the left and without the black data points (and data line) below the clear data points on the right (with the crossover point of the two data plots separating right from left)? Where have you seen this figure before?

While I have your attention, have you thought about Equation 8.2 where the concentration of modifier is approximately half that of the concentration of the enzyme? If there is not a sufficient concentration of modifier to bind to all of the free enzyme, you will have two mechanisms operating for the conversion of substrate to product and your data will look as though the binding of modifier to enzyme is reversible (which is my way of introducing you to Equation 8.3).

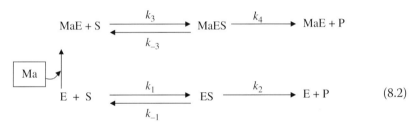

$$(8.2)$$

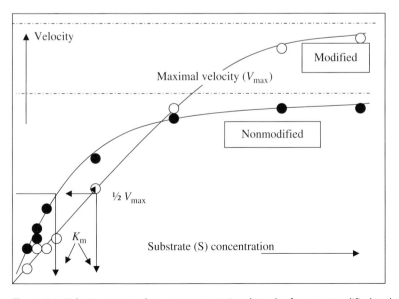

Figure 8.1 Velocity versus substrate concentration data plot for a nonmodified and a modified enzyme to illustrate changes to kinetic constants.

What happens if the binding of the modifier (Ma) to free enzyme (E) is reversible? Let's modify Equation 8.2 to become Equation 8.3. I have indicated the change. I have made binding of the modifier (Ma) to free enzyme (E) reversible. This means that in the presence of modifier (Ma) there will be two mechanisms by which enzyme may convert substrate to product. The problem with Equation 8.3 is that we do not know how much the modified and the nonmodified mechanisms will contribute to the overall conversion of substrate to product. We need rate constants for the binding of modifier to enzyme to produce the modified enzyme (MaE) and for the dissociation of modifier from the modified enzyme (MaE) to re-form free enzyme (E). We also need to emphasize once again, that when describing "in the presence of modifier," we will always mean in the presence of an excess of modifier (unless otherwise specified) where "excess" means there is much more modifier than enzyme present so that modifier concentration does not change during the modification.

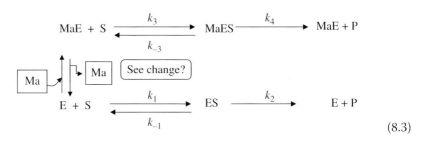

$$(8.3)$$

Let's assign rate constant k_5 for the binding and rate constant k_{-5} for the "unbinding." If we make $k_5 >> k_{-5}$ do you see that we would essentially restore a nonreversible situation? In this case, the dissociation of modifier from enzyme would occur so rarely relative to the association of modifier to the enzyme that for all practical purposes in the presence of sufficient modifier, only the modified mechanism would be available to convert substrate to product. If we make $k_5 << k_{-5}$ we are stipulating that modifier will bind to free enzyme, but the equilibrium of the association/dissociation will be so far toward dissociation that for all practical purposes all substrate conversion to product will occur via the nonmodified mechanism. Do you see why being able to establish relative values for rate constants can be used to stipulate whether or not a modifier will actually modify some enzymatic conversion of substrate to product to such an extent that an experiment will actually reveal modification? Do you understand that the ratio of k_{-5}/k_5 defines an equilibrium binding constant (K_{eq}) such as we discussed earlier for a K_m value for binding of substrate to enzyme under quasi-equilibrium assumptions ($K_m = (k_{-1} + k_2)/k_1$) when $k_2 << k_1$ and/or k_{-1}?

Are you beginning to understand that if I can make up enzymatic mechanisms (modified and not modified) and introduce rate constants where I define how each rate constant relates to all the other rate constants, and produce data plots showing probable velocity versus substrate concentration AND the reciprocal of velocity versus the reciprocal of substrate concentration that you can take your data plots and work backward to try to figure out probable mechanisms for some enzyme reaction that you are studying? The problem will be in trying to decide which of the many many probable mechanisms that you can come up with represent the real (or most appropriate) mechanism. However, you will soon realize that by careful and creative experimental designs you can eliminate some of the possible mechanisms and at least narrow your options down to a few mechanisms.

If you understand this point above, you will understand why I wrote this book. You will also understand why I chose the phrase "Beyond this point there be dragons."

9

Underlying Attributes of Assessing Enzymatic Activities

The greatness of a man is not in how much wealth he acquires, but in his integrity and his ability to affect those around him positively.

Bob Marley

I delayed discussing/presenting the topics to be covered in this chapter until I had the opportunity to cover simple mechanisms for the conversion of substrate to product and the regulation of these conversions by modifiers. This approach worked well in my lectures in that it gave me the opportunity to cover my class in two parts, and students not needing the additional material, to be covered in Part II of this book, could be given lectures directed more to metabolic pathways with a biochemisty emphasis. For those students wishing a greater focus on enzymology, having "part 2" permitted me to continue with more elaborate enzyme mechanisms. Since this book is entitled *Enzyme Regulation in Metabolic Pathways*, you will get the second part in addition to the first part. However, whatever the objective of your education, I would be remiss if I did not at this time touch upon some of the more important attributes associated with the appropriate methods of determining K_m and V_{max} values of enzymes—modified or not modified.

Primary Data Plots

The measurement of enzyme activity is generally relatively simple. One mixes substrate with enzyme and measures the disappearance of substrate or the appearance of product over time. Alternatively, for some enzymatic reactions, other factors (e.g., nicotinamide adenine dinucleotide, or NAD, being converted to a reduced form of nicotinamide adenine dinucleotide, or $NADH + H^+$) may be involved in the enzymatic reaction, and one may observe for some change in these other factors. If the substrate has some specific absorbance spectrum that the product does not have, one can simply measure for a decrease in absorbance at the optimal wavelength of light over time and obtain a graphical representation of the data similar to that of Figure 9.1.

Enzyme Regulation in Metabolic Pathways, First Edition. Lloyd Wolfinbarger.
© 2017 John Wiley & Sons, Inc. Published 2017 by John Wiley & Sons, Inc.

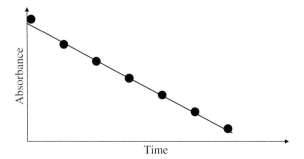

Figure 9.1 A hypothetical data plot showing a decrease in absorbance by substrate over time at a given concentration of substrate and a given concentration of enzyme.

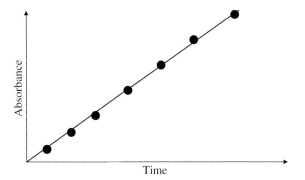

Figure 9.2 A hypothetical data plot showing an increase in absorbance by product over time at a given concentration of substrate and a given concentration of enzyme.

Alternatively, if the product of the enzymatic reaction has some specific absorption spectrum that will permit the monitoring of product, it is perhaps easier to manage the analysis of data and a typical (although hypothetical) data plot obtained at a given concentration of substrate and a given concentration of enzymes would be as illustrated in Figure 9.2.

One reason why it is easy to manage absorbance data, in these two examples (Figures 9.1 and 9.2), is because as we begin to measure changes in absorbance over time at varying concentrations of substrate, the point at which the data line crosses the vertical (*y*) axis will change when measuring changes in absorbance of substrate, but for the most part, product will not be present at zero time for any change in substrate concentration and thus virtually all data plots showing change in absorbance due to appearance of product over time will all originate at the point where the X and Y lines intersect.

The data plot illustrated in Figure 9.2 might seem obvious to even the least knowledgeable biochemistry or biology major. However, it is important for an understanding of the validity of data such as those presented earlier in Figure 3.1, which is presented again in this chapter as Figure 9.4.

Each data point in Figure 9.4 represents the slope of a line such as that shown in Figure 3.2, where each slope value was obtained at a different concentration of

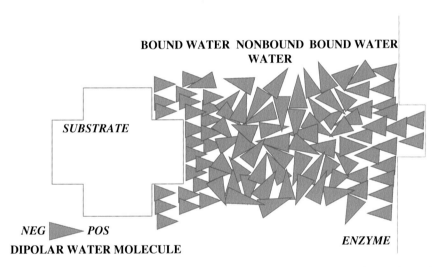

Figure 1.1 Illustration of the role of bound water versus nonbound water in the translational movement of a molecule of substrate into the substrate-binding site of an enzyme.

Enzyme Regulation in Metabolic Pathways, First Edition. Lloyd Wolfinbarger.
© 2017 John Wiley & Sons, Inc. Published 2017 by John Wiley & Sons, Inc.

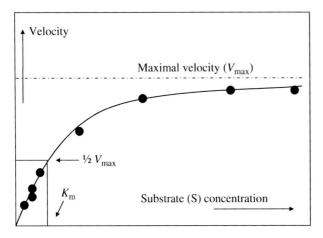

Figure 3.1 Typical velocity versus substrate concentration data plot showing the measured velocity asymptotically approaching maximal velocity at infinitely high concentrations of substrate.

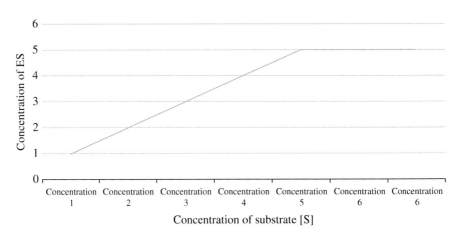

Figure 13.1 Hypothetical graph illustrating the rate of appearance of the "product" (in this case the enzyme–substrate complex as illustrated in Equation 13.1) by plotting the concentration of the enzyme–substrate complex ([ES]) as a function of the concentration of substrate ([S]). Theoretically the line would pass through zero, but this expectation is not shown here simply because the graph is hypothetical, and whether or not the line would pass through zero is speculative at best, but you can think more about that later.

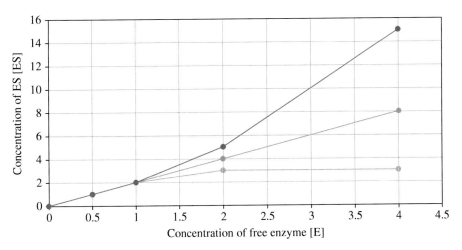

Figure 13.2 A hypothetical graph illustrating the change in the concentration of the enzyme–substrate complex [ES] versus change in the concentration of free enzyme [E] associated with Equation 13.1.

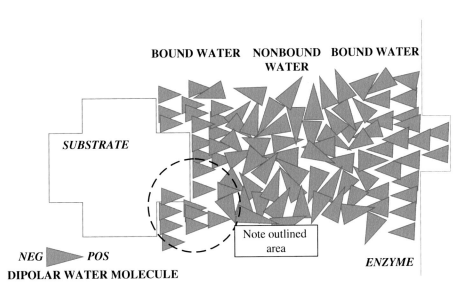

Figure 13.3 Illustration of the role of bound water versus nonbound water in the translational movement of a molecule of substrate into the substrate-binding site of an enzyme [Figure 1.1 repeated].

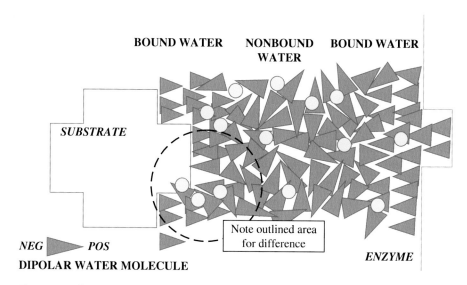

Figure 13.4 Illustration of the role of bound water versus nonbound water in the translational movement of a molecule of substrate into the substrate-binding site of an enzyme where other solute(s) is present to aid in the disruption of the structure of bound water in close proximity to both enzyme and substrate.

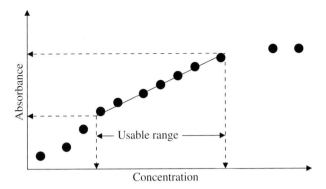

Figure 9.3 A hypothetical data plot showing change in absorbance of some compound, for example "product" as discussed in the text, versus concentration of that compound. As noted, not all absorbance data points will fall on a straight line as determined using regression analysis (normally any one or more data points that cause the R^2 value associated with a regression analysis to become smaller is sufficient reason to exclude it from the regression line derived).

substrate and some common concentration of enzyme. The straight line derived from the individual data points (absorbance values) at each time point is generally obtained using linear regression analysis and is frequently expressed as some change in absorbance over some specified time interval. It is best, however, to convert the absorbance value to some concentration of product based on the extinction coefficient of that product at the prescribed wavelength of light. The extinction coefficient can be determined by measuring the absorbance of increasing concentrations of product (at the specified wavelength of light) and plotting these absorbance values to, hopefully, obtain a straight line. The slope of that line represents the extinction coefficient you will need. Most compounds that exhibit some specific absorbance spectrum will provide for a linear data plot of absorbance versus concentration even though they may not provide for such a linear data plot over a whole range of concentrations of the compound of choice. Such is illustrated in Figure 9.3 where only a short range of concentrations of the compound provide for a linear data plot. You should only use a calculated extinction coefficient (derived from the linear part of such a data plot) to calculate concentrations from absorbance values that fall within the range of the linear part of a data plot such as illustrated in Figure 9.3.

But, let's return to Figure 9.4 (i.e., Figure 3.1 revisited). As stated, each individual data point in Figure 3.1 and in Figure 9.4, represents the slope of a line such as illustrated in Figure 9.2. There are eight such data points in Figure 9.4, but to make Figure 9.5 less busy, I'm going to try to make my point utilizing only every other data point from Figure 9.4, that is, only four slope values that were used originally to gain each data point plotted in Figure 9.4 with each linear regression line (slope) containing four absorbance data points at four different time intervals.

The data presented in Figure 9.5 are of course "ideal" and only obtained when one has figured out the necessary and appropriate range of concentrations of substrate to use in the assessment AND the necessary time frame that will yield

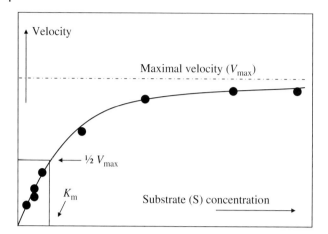

Figure 9.4 Typical velocity versus substrate concentration data plot showing the measured velocity asymptotically approaching maximal velocity at infinitely high concentrations of substrate.

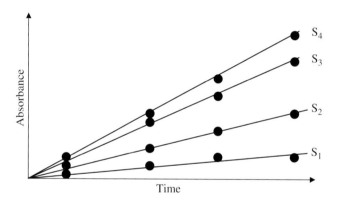

Figure 9.5 A hypothetical data plot showing change in absorbance of product being produced by an enzymatic reaction as a function of time and at four different concentrations of substrate, $[S_1]$, $[S_2]$, $[S_3]$, and $[S_4]$, over four different intervals of time.

a linear change in absorbance over changing time. All too often, inexperienced researchers will chose some range of concentration of substrate that they expect to be appropriate, a time interval over which they expect to obtain a linear data plot of absorbance versus time, and a given concentration of enzyme that may or may not have been pre-calculated based on a, hopefully, known specific activity (usually expressed as units of enzyme per unit weight of protein) for that enzyme preparation. In addition, in order to save time and money, the inexperienced researcher may choose only a given (single) time point at which to measure the change in absorbance (or other measured parameter) assuming that any other data points collected between zero time and the time point chosen will fall on a straight line. This approach can easily lead to major errors in determination of enzyme kinetics, where one obtains the data presented in Figure 9.6, where

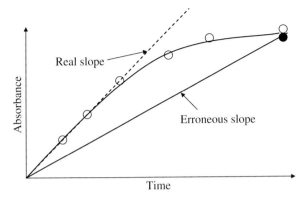

Figure 9.6 A hypothetical data plot showing change in absorbance due to appearance of product over time at some concentration of substrate and a fixed concentration of enzyme. The straight line going from zero to the black circle represents an experiment where a single data point is taken at some fixed time interval. The curved line going from zero to a point equivalent to the data point represented by the black circle represents multiple data points taken at that same concentration of enzyme and substrate. It is clear that the use of a single data point to determine the slope or rate of conversion of substrate to product can lead to an erroneous determination of the rate of conversion if for some reason, the time dependent conversion of substrate to product is not a liner function of time over the time chosen for the experimental determination of rate.

the real slope (i.e., rate of conversion of substrate to product) is erroneously calculated using a single data point.

The nonlinearity of the data plot in Figure 9.6 over the time frame being studied can be for a variety of reasons. For example, the concentration of substrate could be so low (or the enzyme concentration so high) that the concentration of substrate is significantly changing during the course of the assessment. If the concentration of substrate decreases to such an extent that it impacts on the equilibrium concentration of the enzyme–substrate (ES) complex (under quasi-equilibrium assumptions) by reducing the association of substrate and free enzyme, then the measured velocity will decrease proportional to the decrease in substrate concentration, and the researcher will obtain the results illustrated in Figure 9.6. Alternatively, if the overall conversion of substrate to product is reversible, then as product begins to accumulate, the rate of conversion of product to substrate may cause the absolute concentration of product (P) to plateau and one will obtain the data plot illustrated in Figure 9.6. Finally, one may also obtain the data plot illustrated in Figure 9.6 if product feedback inhibits the enzyme by either increasing its K_m or by decreasing its V_{max} value through some modified enzymatic mechanism. Such would be characteristic of, for example, the enzyme converting phenylpyruvic acid to phenylalanine in the aromatic amino acid biosynthetic pathway.

In short, in obtaining data for primary data plots and calculation of velocities of conversion of substrate to product at variable concentrations of enzyme or substrate, it is essential that the investigator insures that he/she is obtaining a rate of conversion of substrate to product that represents a linear function over time.

Secondary Data Plots

I utilized the familiar Lineweaver–Burk data plot (see Figure 3.2) to illustrate how data plotted from velocity versus substrate concentration data plots (see Figure 3.1) could be used to better visualize enzyme mechanisms (whether non-modified or modified by modifiers). However, there are other methods of plotting data that are analogous to the Lineweaver–Burk data plots that you should be familiar with. Figure 9.7 illustrates the Eadie–Hofstee plot useful for replotting of data from a standard velocity versus substrate concentration data plot.

Although the Eadie–Hofstee plot is typically thought of as being more difficult to easily convert the values of velocity versus substrate concentration to velocity versus velocity divided by the substrate concentration, the Eadie–Hofstee plot has the advantage over the Lineweaver–Burk data plot of "normalizing" the velocity of an enzymatic reaction to the substrate concentration and tends not to bias slope determinations by giving "equal weight" to velocity values obtained at the lower substrate concentrations, which tend to present more error than the greater velocity values at the higher substrate concentrations as found with the Lineweaver–Burk data plots. By dividing each measured velocity by the substrate concentration used to obtain that velocity value, data points at lower substrate concentrations present less bias and thus the Eadie–Hofstee data plots are generally considered to be superior to Lineweaver–Burk data plots.

Let's see how data resulting from modified and nonmodified enzymatic mechanisms will look when plotted using Eadie–Hofstee data plots. Figure 9.8 shows the Eadie–Hofstee method of plotting velocity versus substrate concentration data when the enzyme conversion of substrate to product occurs in the presence of competitive modifiers acting as either an inhibitor or an activator. It results in the data plots crossing the vertical (y) axis at the same point as the nonmodified enzyme—meaning competitive modifiers do not change the apparent V_{max} values of the enzyme. However, the points of intersect on the x axis (i.e., velocity/

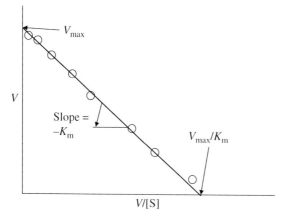

Figure 9.7 The Eadie–Hofstee plot. "V" stands for velocity, such as would be obtained from data like those shown in Figures 9.1 and 9.2, and [S] stands for substrate concentration.

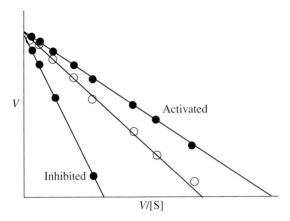

Figure 9.8 Eadie–Hofstee data plot showing velocity versus velocity/substrate concentration values for nonmodified (open circles) enzyme and for modified enzymes being activated or inhibited "competitively" by some modifier (black circles).

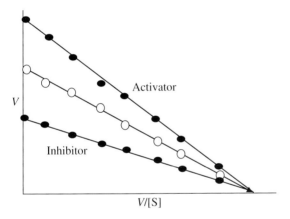

Figure 9.9 Eadie–Hofstee data plot showing velocity versus velocity/substrate concentration values for nonmodified (open circles) enzyme and for modified enzymes being activated or inhibited "un-competitively" by some modifier (black circles).

[S]) will move to either the left (inhibitor) or to the right (activator) of the point of intersect of the nonmodified enzyme. This would seem obvious in that if for a given concentration of substrate one achieved a lower measured velocity in the presence of a modifier, the value of $V/[S]$ should get smaller at each substrate concentration if the modifier were an inhibitor and larger if the modifier were an activator. That the slope of the line in such data plots changes would be consistent with a competitive modifier changing the apparent K_m of the enzymatic reaction. The steeper the slope of the data plot, the bigger the apparent negative K_m.

"Uncompetitive" modifiers change the apparent V_{max} and the apparent K_m values of enzymes. So if we plot data obtained for enzyme in the presence of "uncompetitive" modifiers (modifiers acting as activators or inhibitors), we will obtain data such as shown in Figure 9.9.

The Eadie–Hofstee method of plotting velocity versus substrate concentration data when the enzyme conversion of substrate to product occurs in the presence of an uncompetitive modifier acting as either an inhibitor or an activator will result in the data plot crossing the x (horizontal) axis at the same point as the nonmodified enzyme—meaning uncompetitive modifiers change both the apparent K_m and V_{max} values of the enzyme. However, the points of intersect on the vertical (y) axis (or the velocity) will move either down (inhibitor) or up (activator) relative to the point of intersect of the nonmodified enzyme. This would seem obvious in that if for a given concentration of substrate one achieved a lower measured velocity in the presence of a modifier, the velocity of the enzymatic reaction should get smaller at each substrate concentration if the modifier were an inhibitor and larger if the modifier were an activator. That the slope of the line in such data plots changes would be consistent with an uncompetitive modifier changing both the apparent V_{max} and K_m of the enzymatic reaction. The steeper the slope of the data plot, the bigger the apparent negative K_m.

The data plot in Figure 9.10 illustrates the kind of data plot you will obtain when plotting noncompetitive modifiers using the Eadie–Hofstee representation. You will note immediately that the intercepts on both the x and y axes for the modified enzyme data plots (black circles) change relative to the nonmodified enzyme data plot (open circles). This is because a noncompetitive modifier changes the apparent V_{max} of the enzyme, but not the apparent K_m of the enzyme (the slope does not change relative to the nonmodified enzyme data plot).

By now, it should be obvious to you that one advantage the Eadie–Hofstee data plot has over the Lineweaver–Burk data plot has to do with the positions of individual data points on both kinds of data plots. With the Eadie–Hofstee data plots, modifiers acting as activators "always" appear with the modified data points above the same data points in the nonmodified enzyme assay, and

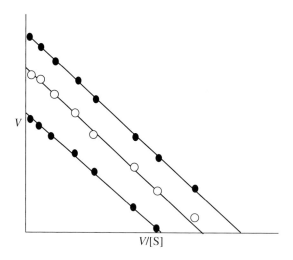

Figure 9.10 Eadie–Hofstee data plot (noncompetitive) showing velocity versus velocity/substrate concentration values for nonmodified (open circles) enzyme and for modified enzymes being activated or inhibited "noncompetitively" by some modifier (black circles).

inhibitors "always" appear with the modified data points below the same data points in the nonmodified enzyme assay. With the Lineweaver–Burk data plots, modifiers acting as activators "always" appear with the modified data points below the same data points in the nonmodified enzyme assay, and inhibitors "always" appear with the modified data points above the same data points in the nonmodified enzyme assay. In this respect, the Eadie–Hofstee data plot is easier to understand than the Lineweaver–Burk data plot, and we will consider the use of "always" in later chapters.

Part II

Well, I would like to be a little larger, Sir, if you don't mind," said Alice. "Three inches is such a wretched height to be." "It is a very good height indeed," said the caterpillar angrily...

Lewis Carroll, *Alice in Wonderland*

10

Breakdown of the Michaelis–Menten Equation (or Complex Enzyme Mechanisms)

Because I did not wish to involve you in enzyme-driven chemical reactions involving more than one substrate in Part I of this book, I restricted my description of enzyme mechanisms to single substrate/single product reactions. This approach kept things simple and let me focus on getting you to figure out how you could manipulate rate constants and their relationships to each other to come up with mechanisms that would support any data you might obtain experimentally.

However, these earlier chapters dealing with certain specifics of how data can be (should be) collected and analyzed afforded me the opportunity to consider how you might deal with what I will call "multi-substrate enzyme reactions" before going onto more complex enzyme mechanisms. Obviously, this section can become complicated and yield a huge number of equations and correspondingly complicated data plots. Since this textbook is intended to cover how enzyme modification(s) are involved in regulating the flow of metabolites through metabolic pathways, I chose to keep this description of multi-substrate enzyme reactions as simple as possible. The most simple mechanism would be to have an enzyme involved in the conversion of two substrates into a single product such as illustrated in Equation 10.1.

$$E + A \rightleftharpoons EA \;\; \underset{B}{\overset{B}{\rightleftharpoons}} \;\; EAB \longrightarrow E + P \qquad (10.1)$$

Although Equation 10.1 tends to suggest that A binds first to free enzyme (E) to form an EA complex and then B binds to the EA complex to form the EAB complex before breaking the EAB complex to free enzyme (E) and product (P), the equation is not intended to imply an ordered sequence of binding of "substrates" to free enzyme. One could easily allow substrate B to bind first to free enzyme and for substrate A to bind to an EB complex to form the EBA complex.

Either way, this "multi-substrate" enzyme-driven reaction is difficult to analyze in that there will be two changing substrate concentrations to deal with, and the sequence of binding of substrates may be important. Typically, solutions to these kinds of enzyme-driven chemical reactions have involved adding sufficient concentrations of one substrate (e.g., A) to "saturate" the enzyme reaction, and varying the concentration of the other substrate (e.g., B), while determining the disappearance of one (or both) of the substrates or appearance of product over time. You then add sufficient concentrations of the other substrate (B) to saturate the enzyme reaction

Enzyme Regulation in Metabolic Pathways, First Edition. Lloyd Wolfinbarger.
© 2017 John Wiley & Sons, Inc. Published 2017 by John Wiley & Sons, Inc.

with B and vary the concentration of the other substrate (A), and again measure the disappearance of one (or both) of the substrates or appearance of product over time. Obviously, you will probably have to repeat these studies until you are able to determine what the saturating concentrations of substrates A and B really are, but much of research involves a learning curve with regards to experimentation.

Of course there is another way of expressing a random, but sequential order mechanism for the conversion of two substrates to one product, as illustrated in Equation 10.2

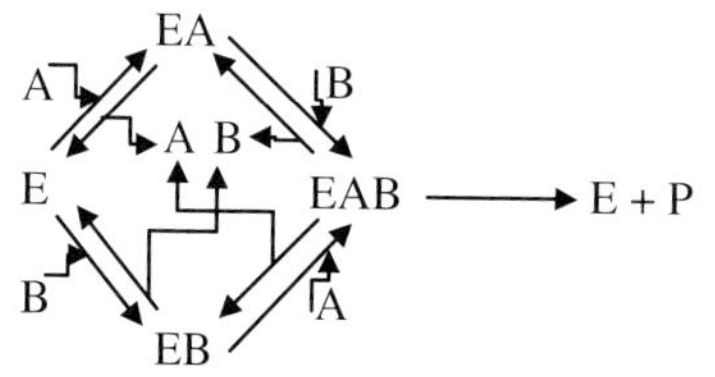

$$
\text{(10.2)}
$$

Equation 10.2 is essentially the same as Equation 10.1, but better illustrates the various options for which substrate binds first and which substrate binds second. Should the sequence of substrate(s) binding be determined to be an ordered sequential mechanism, it is more appropriate (easier) to use Equation 10.1 when applying rate constants to the mechanism.

It is easy to imagine how complicated this kind of enzyme-driven chemical reaction can become should it be a two-substrate and two-product mechanism such as is illustrated in Equation 10.3.

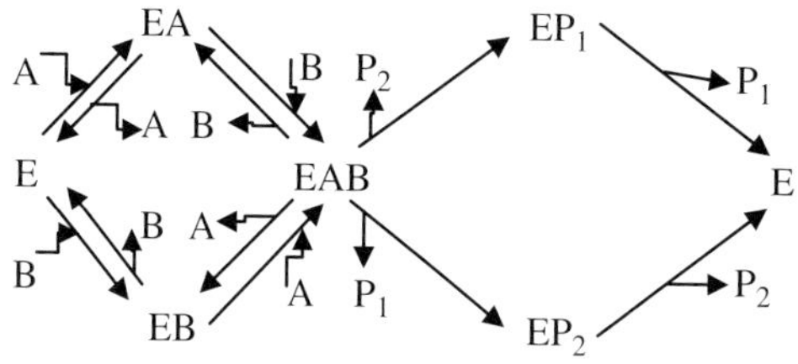

$$
\text{(10.3)}
$$

Equation 10.3 implies, of course, that there is a random order and sequential binding of substrates, and a sequential and random order release of products, but that the formation of products (P_1 and P_2) is not reversible, similar to the product formation in Equation 10.2.

You noticed that rate constants were not provided in Equations 10.1, 10.2, and 10.3. The idea for these first three equations in this chapter was simply to get you to begin visualizing how you can express enzyme mechanisms where there is more than one substrate or more than one product.

Before we get into more complicated mechanisms, let's go back to Equation 10.1 and provide (organize) some rate constants for each step in the mechanism. This will be give us Equation 10.4.

$$
E + A \underset{k_{-1}}{\overset{k_1}{\rightleftharpoons}} EA \underset{\substack{B \\ k_{-2}}}{\overset{\substack{B \\ k_2}}{\rightleftharpoons}} EAB \overset{k_3}{\longrightarrow} E + P
$$

$$
\text{(10.4)}
$$

Be careful now and appreciate that I am about to make k_3 the rate-limiting step in this mechanism rather than the k_2 that you were familiar with in Part I of this book.

The question becomes, how do we wish to establish relative values for the rate constants in this mechanism? We can begin by establishing quasi-equilibrium assumptions in this mechanism by letting k_3 be much much smaller than all the other rate constants. This would mean that the conversion of EAB to $E + P$ is the rate-limiting step in the conversion of two substrates to one product by this mechanism. We can then make the mechanism pretty simple by letting k_1 equal k_2 and k_{-1} equal k_{-2}. This relationship would mean that substrate A associates and dissociates to/from free enzyme equal to the way substrate B associates and dissociates to/from the EA complex (the "enzyme" binds both substrates equivalently). We thus have an ordered sequential two-substrate, one-product, enzyme reaction where the rate-limiting step k_3 is not reversible.

Have you guessed the next step in this process? What if we have a compound that will bind to free enzyme and modify the overall conversion of two substrates to one product as either an "inhibitor" or an "activator"? And you didn't think you knew where this was going! Let's start out simple. Let's look at a mechanism where there is only one modifier acting on the free enzyme. Also, let's start with this modifier being an "inhibitor" of the enzymatic conversion of two substrates to one product. Let's look at Equation 10.5.

$$
\begin{array}{ccccccc}
& k_1 & & \overset{B\rceil}{} \ k_2 & & k_3 & \\
E + A & \underset{k_{-1}}{\overset{}{\rightleftharpoons}} & EA & \underset{B\lfloor k_{-2}}{\overset{}{\rightleftharpoons}} & EAB & \longrightarrow & E + P \\
\Big\downarrow Mi & & & & & & \\
& k_4 & & \overset{B\rceil}{} \ k_5 & & k_6 & \\
MiE + A & \underset{k_{-4}}{\overset{}{\rightleftharpoons}} & MiEA & \underset{B\lfloor k_{-5}}{\overset{}{\rightleftharpoons}} & MiEAB & \longrightarrow & E + P
\end{array} \qquad (10.5)
$$

By now, you know that Equation 10.5 is not complete because we have not established any relationships between the rate constants. No rate constant has been assigned for the binding of the modifier (Mi) to free enzyme, but since the binding is designated as not reversible, we will stipulate that in the presence of excess concentrations/amounts of modifier, when modifier is present, all enzyme-mediated conversion of two substrates to one product will occur via the modified enzyme mechanism (as in Equation 10.5).

What now can be more simple in designating Mi as an inhibitor of this enzymatic conversion of two substrates to one product than to stipulate that k_6 is less than k_3 ($k_6 < k_3$) and that this inhibitor will be a *noncompetitive* inhibitor, changing the apparent V_{max}, but not changing the apparent K_m of the enzyme-mediated reaction. How strong an inhibitor Mi will be depends, of course, on how you might stipulate how much or how little "less" k_6 might be relative to k_3. If $k_6 << k_3$, then the modifier (Mi) could be construed as being similar to the use of boiling water to inactivate all of the enzyme.

However, we now have two options for Mi to be a *competitive* inhibitor of this enzyme-mediated conversion of two substrates to one product. We can let the modifier change the association constants for the binding of either substrate A or substrate B to their appropriate enzyme (either free enzyme E or substrate-bound free enzyme EA). Either option will permit designation of the modifier (Mi) as a competitive inhibitor, but in the *normal mode* (traditional teachings) of describing a competitive inhibitor as being a molecular analog of a substrate and binding in the same site as a substrate (which of course we will use but not accept as the only option), we would mostly be determining which substrate the modifier was an analog (similar structure) of, substrate A or substrate B. Either way, the modifier Mi would change the apparent K_m of either substrate A or substrate B, but not the maximal rate of conversion of the two substrates to one product since we would, of course, stipulate that k_3 equaled k_6 ($k_3 = k_6$).

In that I rarely operate in "normal mode" with respect to enzyme kinetics, let's assume the following: that k_4 is less than k_1 ($k_4 < k_1$) AND that k_2 is less than k_5 ($k_5 < k_2$), but k_3 equals k_6 ($k_3 = k_6$) and that $k_4 = k_5$. This relationship of rate constants would be similar to earlier rate constant relationships, where the modifier was a molecular analog of one of the substrates, in providing for the modifier (Mi) to be a competitive inhibitor of the conversion of two substrates to one product by modifying the apparent K_m of the enzyme reaction, but not modifying the apparent V_{max} of the enzyme reaction. However, in this case, the modifier (Mi) would be changing the association constants of both substrates to their respective forms of the enzyme to some lesser values. This mechanism would tend to suggest that the modifier (Mi) was bound at some site other than either of the two substrate-binding sites and in so doing changed the rates at which substrates A and B bound to their respective substrate-binding sites. We would have a "competitive" modification of the enzyme reaction, but it would not be what you might call a "normal" mode of competitive inhibition, and any data plot such as a Lineweaver–Burk data plot would look like what was illustrated earlier in Figure 5.1. Remember, according to Equation 10.5, binding of the modifier to free enzyme is not reversible—there is more to come on this topic.

Let's expand on Equation 10.5 and create a new mechanism, which we will label as Equation 10.6.

$$(10.6)$$

Are you having fun yet? Looking at Equation 10.6, what we have is an enzyme mechanism for converting two substrates (A and B) to one product (P) via one nonmodified enzyme mechanism and via two modified enzyme mechanisms. One of the modified enzyme mechanisms is modified as in Equation 10.5 by an inhibitor (Mi), but now we have a second modified enzyme mechanism where the enzyme is modified by a modifier we have so far been describing as an "activator" (Ma). So, it would appear that what we have in Equation 10.6 is an ordered sequential two-substrate one-product nonreversible enzyme mechanism that can be both inhibited AND activated by the appropriate modifiers, and that modifier binding is nonreversible. However, we have not established relationships between the various rate constants and thus we don't know what kinds of data plots we will obtain when we measure enzyme activity in the absence and/or presence of (both?) modifiers.

I suggest that we keep this simple and simply stipulate that $k_1 = k_2$, that $k_{-1} = k_{-2}$, that $k_1 \geq k_{-1}$, that $k_2 \geq k_{-2}$, and that k_1 (and therefore k_2) $>> k_3$ for the nonmodified mechanism. Then we should stipulate that $k_4 = k_5$, that $k_{-4} = k_{-5}$, that $k_4 \geq k_{-4}$, that $k_5 \geq k_{-5}$, and that k_4 (and therefore k_5) $>> k_6$ for the mechanism associated with the Mi-modified mechanism.

This then leaves us with only the modified mechanism associated with the Ma modifier, and if we let the rate constants in this modified mechanism be that $k_7 = k_8$, that $k_{-7} = k_{-8}$, that $k_7 \geq k_{-7}$, that $k_8 \geq k_{-8}$, and that k_7 (and therefore k_8) $\gg k_9$ we will have all three enzyme mechanisms operating under quasi-equilibrium assumptions.

If we deal with the enzyme mechanism associated with the Ma modifier and stipulate that k_9 is greater than k_3, we will have a modified enzyme mechanism that in the presence of modifier Ma (and conversely in the absence of modifier Mi), represents an activation where the conversion of the two substrates to one product in the presence of Ma will occur at a faster rate (greater velocity) than in the absence of modifier Ma.

The problem we have, however, is that either modifier (Ma or Mi) can divert all of the free enzyme to either MaE or MiE (given the usual conditions), and we thus have a rather uninteresting situation where we will just shift from one to another modified enzyme mechanism depending on which modifier is present "first." This situation is just plain boring. Let's spice the situation up a bit. Let's make the binding of modifiers reversible!

But first, let's add another option to the overall mechanism. Let's add a second modifier to the options. If we stipulate that the Ma modifier can actually be present as either Ma1 or Ma2—meaning two distinct modifiers that will have the capacity to activate the free enzyme to convert the two substrates to one product at a greater rate (increased velocity)—we can add significantly to the next equation. Equation 10.7 is going to be really interesting. I'll bet you are really excited by now.

What we now have in Equation 10.7 is a two-substrate one-product ordered sequential enzyme reaction that can be reversibly modified by three modifiers (at least three known modifiers for now) giving a total of four enzyme mechanisms available for the conversion of the two substrates to the one product. Knowing that the "Force is with you," I'm going to simply stipulate that all four enzyme mechanisms operate under quasi-equilibrium assumptions (you can figure out the relationships of the rate constants in each mechanism since we have already done this several times now). In addition, we can anticipate that since two of the modifiers are "activators," that the corresponding rate-limiting rate constants (k_9 and k_{12}) will be greater (at least for now) than the rate-limiting rate constant k_3 in the nonmodified mechanism, AND that the one modifier designated as an "inhibitor" will provide for a modified enzyme mechanism where the rate-limiting rate constant k_6 will be smaller than the rate-limiting constant k_3 for the non-modified enzyme mechanism. Again, I am taking a simple option such that each modifier is changing only the apparent V_{max} value in each modified mechanism (and not modifying the apparent K_m values of these modified mechanisms). Equation 10.7 looks really complicated, but if you walk yourself through it one step at a time with me it will become clearer.

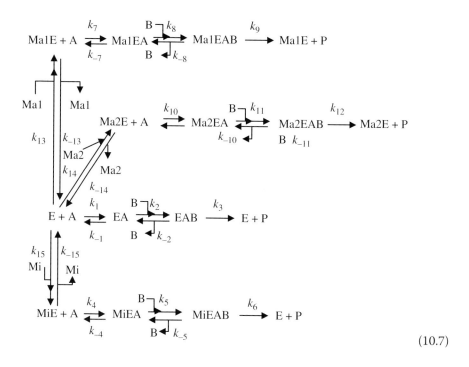

$$(10.7)$$

Looking at the whole of the total enzyme mechanisms illustrated in Equation 10.7 you can determine that we have a nonmodified enzyme mechanism for the enzymatic conversion of two substrates to one product where the addition of substrates to the free enzyme is ordered (meaning A before B) and sequential (one after the other) and that the formation of product P is not reversible.

In the presence of modifiers (and there are three possible in this mechanism), each of three modifiers may bind to free enzyme forming three different forms of modified enzyme, and these bindings of modifiers to free enzyme are reversible.

You can imagine that in the presence of all three modifiers, some combination of all four enzyme mechanisms will be converting the two substrates to the one product and how, at some equilibrium, the relative concentrations of enzyme functioning in each of the four mechanisms will depend mostly on the relative concentrations of each modifier (or possibly the concentration of the two substrates—but that comes later) in that we have stipulated that the modifiers change only the apparent V_{max} values of each mechanism. Should one modifier become lower in concentration and the modified form of the enzyme shift back to the nonmodified form of the enzyme, you can expect that the whole equilibrium situation will readjust in the relative amounts of enzyme available to convert the two substrates to the one product via each of the four mechanisms. Given a constant and stable amount of enzyme being produced via transcription of the appropriate gene and translation of the transcribed messenger RNA into protein with subsequent post-transcriptional modification of that protein to a functioning enzyme, we should expect a given cell possessing this metabolic capability to achieve some steady-state level of production of product P (or conversely the consumption of the two substrates A and B).

I'll bet you know by now where we are going next. You are correct, let's look again at Figure 4.4, which we will renumber as Figure 10.1, so you don't have to flip back through the book.

I will remind you that I shortened the aromatic amino acid biosynthetic pathway in Figure 4.4, and now again in Figure 10.1, in that I could not get the whole pathway into a figure. The three isozymes shown as enzymes "a," "b," and "c" are three isomeric forms of the enzyme D-arabinoheptulosonate-7-phosphate synthetase (or DAHP synthetase); this enzyme converts the two substrates erythrose-4-phosphate and

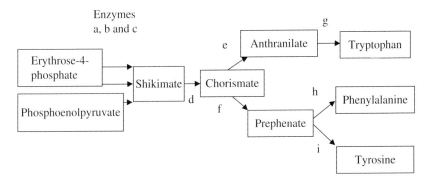

Figure 10.1 Pathway as previously shown in Figures 4.3 and 4.4, with specific molecules found in this aromatic amino acid biosynthetic pathway substituted for the symbols used in Figure 4.3. Prephenate is actually metabolized to either phenylpyruvate prior to formation of phenylalanine, or to parahydroxyphenylpyruvate prior to formation of tyrosine, but these intermediates were left out of this scheme in order to emphasize the enzyme steps relevant (important) to regulation of this metabolic pathway and to aid in comprehension.

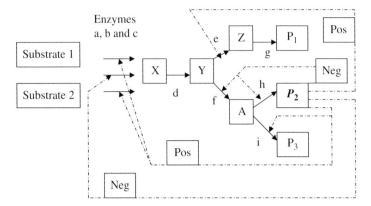

Figure 10.2 Diagram of the aromatic amino acid biosynthetic pathway showing the enzymatic steps in the pathway at which phenylalanine (P₂) modifies enzymatic activity via positive and negative regulatory events.

phosphoenolpyruvate to one product, D-arabinoheptulosonate-7-phosphate (not shikimate as illustrated—this was one of the shortenings of the pathway).

Because I do not wish to end up with a short rope around my neck and a long drop under my feet, I am once again going to take this effort further by focusing on one of the three isozymes, but feel free to apply what I discuss below as relevant to the other two isozymes—just with different "modifiers." I have again reproduced an earlier figure (Figure 4.6) and renumbered it as Figure 10.2 so you don't have to flip back through the text.

As you can see in Figure 10.2, the end product P_2 (which is designated as phenylalanine in this figure) exerts negative feedback inhibition of what I have designated as "enzyme b" (at least according to the figure). If you look back at Figures 4.5 and 4.7, you will see that the other two "end products" (tyrosine and tryptophan) of the aromatic amino acid biosynthetic pathway exert a positive feedback activation of what I have designated in Figure 10.2 as "enzyme b."

Appreciate that just now I am taking a very narrow perspective of the aromatic amino acid biosynthetic pathway, focusing on only one enzyme in that highly regulated pathway. But for now, I need to remain focused or I will lose the emphasis it has taken almost 10 chapters to gain.

It is now time for you to begin threading all of this information into an understanding of how complicated the regulation (only at the enzyme level in this book) of intermediary metabolism can be. In just the simple regulation of one isozyme ("enzyme b") in one metabolic pathway, there are at least four (there are more here, but we will ignore this for ever) possible enzyme-mediated mechanisms for the conversion of two substrates (erythrose-4-phosphate and phosphoenolpyruvate) to one product (D-arabinoheptulosonate-7-phosphate). One of these four mechanisms would be a nonmodified enzyme mechanism; one of these four mechanisms would be a modified enzyme mechanism where the modifier exerts an "inhibitory" effect on the enzyme; and two of these four mechanisms would be a modified enzyme mechanism where the modifier exerts an "activation" effect on the enzyme. Now, look again at Equation 10.7. Don't look back in the book; I will reproduce it again just below this paragraph for emphasis.

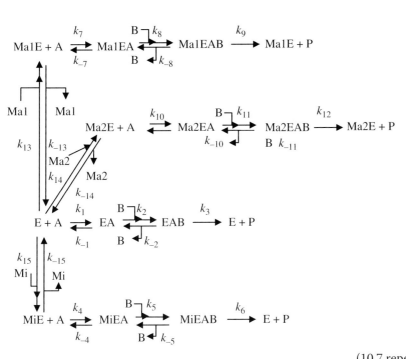

(10.7 repeated)

The possible enzyme mechanisms for the enzyme-mediated chemical conversion of erythrose-4-phosphate and phosphoenolpyruvate to D-arabinoheptulosonate-7-phosphate by the enzyme D-arabinoheptulosonate-7-phosphate synthetase (in this case "enzyme b" of the three available isozymes of this synthetase) can be illustrated in Equation 10.7. You will note that I have said "possible" because we do not know what the actual mechanisms of this enzyme are, nor do we know the actual mechanisms by which it is regulated. However, once again, it is my objective to get you to begin applying yourself to understanding how you might think about such matters rather than providing you with answers.

I left you in the text above with the regulatory events associated with both the inhibition (by phenylalanine) and the activation (by both tryptophan and tyrosine) of this synthetase being mediated solely by modification of the apparent V_{max} values of each of the three modified enzyme mechanisms. I would be remiss in my duties if I left you with such a simplistic view, and thus I feel I must continue with my description of possible regulatory mechanisms for this synthetase.

We have already covered how modification (change) of the apparent K_m value of an enzyme can result in lesser or greater rates of conversion of substrate to product without changing some apparent V_{max} of that same enzyme. However, this approach is a great deal more complicated than simply changing apparent V_{max} values of multiple possible enzyme mechanisms through action(s) of some modifier (especially some modifier that is associated with the metabolic pathway in which that enzyme is functioning). Let's approach this mode of regulation/modification in a logical manner. Let's hold the V_{max} values of all modified and nonmodified enzymes constant and look only at how changing the apparent K_m values of that enzyme will impact on the rate(s) of conversion of substrate to product.

I have already suggested to you that the concentration of a given substrate within a cell will approximate the K_m value of "the" enzyme that is converting it to some other compound. Yes, I know that any given substrate is probably metabolized by multiple enzymes, but for now bear with me. This intracellular concentration of a "substrate" means that as the concentration of that substrate declines, the rate of its conversion to product will change (decrease) in an almost linear fashion (look at the velocity vs substrate concentration curve in Figure 3.1). Conversely as the concentration of that substrate increases, the rate of its conversion to product will change (increase) in an almost linear fashion. This almost linear change means that the cell has an extremely sensitive means of controlling metabolic activity centered around any and all enzymes that maintain intracellular concentrations of metabolites at or near to the K_m values of the enzymes that convert them to other compounds. Imagine, for example, that if the concentration of some substrate was at or near to the "saturating" concentration for the enzyme metabolizing it. Sudden increases or decreases in the concentration of that metabolite would little change its enzymatic conversion to some product in some metabolic pathway.

We thus have another way, other than changing the apparent V_{max} of some modified enzyme mechanism, to shift the flow of metabolites in intermediary metabolism using those very same metabolites. We have enabled some metabolite intermediate to alter the apparent K_m of the overall enzyme conversion of a substrate to a product and in so doing we will find that our options for *how* we alter that apparent K_m are considerable.

To begin our options discussion, let's start with just a small part of the overall enzyme mechanism shown in Equation 10.7 (and 10.7 repeated). Let's think about just that part of Equation 10.7 that is "boxed" in Equation 10.8. You will notice that no product (P) formation lies inside the "box" in Equation 10.8 and we are focused only on the unmodified enzyme mechanism and one (of the two) modified enzyme mechanisms (where Ma2 is the modifier of free enzyme); the intent here is to examine the role of an "activator" in modification of one (or both) of the K_m values that either substrate A or substrate B will provide when associating and disassociating with an enzyme (whether that enzyme be E, EA, Ma2E, or Ma2EA).

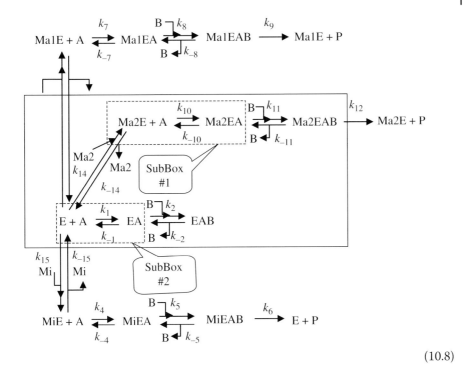

$$(10.8)$$

Equation 10.8 is provided to illustrate a small part of Equation 10.7 (that part where the association and dissociation of substrate A binds to either free E or free Ma2E. There are also two "subboxes" labeled "SubBox #1" and "Subbox #2" to delineate and separate the two mechanisms of interest at this time, showing just those rate constants that will be involved in the apparent K_m of these two mechanisms. Remember we are operating under quasi-equilibrium assumptions.

To deal with even this small part of the overall enzyme mechanisms we developed as Equation 10.7, which we associated with just one of the three isozymes (enzyme b) as illustrated in Figure 10.2, we must continue to focus our attention on fewer and fewer parts of the equation to a point where we focus on just two reversible parts of two mechanisms (one not modified and one modified), specifically the association and dissociation of substrate A with free enzyme E, AND the association and dissociation of substrate A with free modified enzyme Ma2E. Since we are operating under the assumption that rate constants k_3 and k_{12} are negligible in any respective K_m value for these two mechanisms (and that we are not dealing at present with the binding of the second substrate, B), we may define the K_m values for this equilibrium binding of substrate A to its respective free enzymes as:

$$K_m = k_{-1} / k_1 \left(\text{for substrate A to bind to E} \right) \left(\text{SubBox \#2} \right)$$

$$K_m = k_{-10} / k_{10} \left(\text{for substrate A to bind to Ma2E} \right) \left(\text{SubBox \#1} \right)$$

Our interest in the K_m value for k_{-1}/k_1 is only that it represents the K_m of the enzyme-mediated conversion of substrate to product for the nonmodified enzyme mechanism, and if we establish any rate constants for a modified enzyme that are different from any rate constants in the nonmodified enzyme mechanism, we will have changed the apparent K_m for the overall conversion of substrate to product.

Since we are working with what we have described as an activation of the enzyme conversion of substrate to product, let's take a first step and stipulate that k_{10} is greater than k_1 ($k_{10} > k_1$), but that k_{-1} equals k_{-10} ($k_{-1} = k_{-10}$). At this point, substrate A will bind to free enzyme Ma2E faster than it will bind to free enzyme E, but will dissociate from either form of free enzyme at the same rate. All things being equal (*we are not done yet*) at this point, in the presence of the modifier Ma2, more substrate A will be bound to free enzyme Ma2E to form a greater concentration of Ma2EA than will be bound to free enzyme E to form a given (*we are not done yet*) concentration of EA. If we did not have to consider the binding of substrate B, and had stipulated that the rate-limiting steps in the modified and nonmodified enzyme mechanisms were the same, we would note a greater rate of conversion of substrate(s) to product in the presence of modifier Ma2 at virtually all substrate concentrations until such substrate concentrations began to approach saturation of the enzyme with substrate (i.e., V_{max} would not change, but velocities less than V_{max} will change with increasing concentrations of substrate).

However, let's also think about what we just did to the K_m value for the modified enzyme mechanism for binding substrate A to free enzyme Ma2E. Let's first isolate our "definitions" of K_m from the text as we did above.

$$K_m = k_{-1} / k_1 \left(\text{for the nonmodified mechanism in SubBox \#2}\right)$$

$$K_m = k_{-10} / k_{10} \left(\text{for the modified mechanism in SubBox \#1}\right)$$

Let's play with some real numbers for these two respective rate constants (k_{-1} as the dissociation constant and k_1 as the association constant) for the nonmodified mechanism. Let's assign real numbers to these two rate constants: let $k_1 = 10{,}000$ and $k_{-1} = 5000$. These real numbers will provide us with a more "real" K_m value simply by inserting the "real numbers" into our equation of K_m:

$$K_m = k_{-1} / k_1 = 5000 / 10{,}000 = 0.5$$

Let's do the same thing for the two respective rate constants (k_{-10} as the dissociation constant and k_{10} for the association constant) for the modified mechanism. Let's assign real numbers to these two rate constants; let $k_{10} = 100{,}000$ and $k_{-10} = 5000$. As with the nonmodified enzyme rate constants described

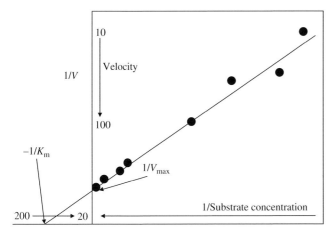

Figure 10.3 A typical Lineweaver–Burk data plot that plots the reciprocal of the velocity against the reciprocal of each substrate concentration. The values of V_{max} and K_m are calculated by extrapolation of where the data line crosses the *x* and *y* axes for 1/Velocity and 1/Substrate concentration. Note that I have applied "relative" numbers for velocity of reaction along the *y* axis (10 to 100 downwards) and 200 to 20 left to right on the negative side of the *x* axis to illustrate the essential complications of a double reciprocal data plot.

above, we can let these real numbers provide us with a more "real" K_m value simply by inserting the "real numbers" into our equation of K_m:

$$K_m = k_{-10} / k_{10} = 5000 / 100{,}000 = 0.05$$

This exercise is not about reality, but rather about getting to a change to an apparent K_m value that is one order of magnitude different, that is, 0.5 versus 0.05. The reality of this exercise, however, is also that you would not see such an order of magnitude difference because there will still be enzyme conversion of substrate to product by the nonmodified enzyme mechanism; but for now just focus on the one order of magnitude difference I am using to illustrate a point.

I have already said that a K_m value was "equivalent to that substrate concentration…" and we may therefore think of the 0.5 and 0.05 as some "concentration." Let's think about substrate concentrations in millimolar (mM) values. If we do this conversion, we end up with K_m values of 0.5 mM and 0.05 mM. Which of these two values represents a greater concentration and a lesser concentration of the substrate that will be necessary to achieve a measured velocity of the conversion of substrate to product at one-half of the maximal rate possible (V_{max})? It should be obvious to you that the modified enzyme mechanism will now bind substrate A to free enzyme Ma2E at a concentration of substrate A equal to 0.05 mM at one-half V_{max}, and that the free enzyme E will require a concentration of substrate A equal to 0.5 mM at one-half V_{max}, that is, one order of magnitude concentration lower.

In that I'm pretty sure you got this point, I want to make one additional point that you may need to flip back to Figure 3-2 to see, so I will repeat Figure 3.2 here as Figure 10.3. If you wish to convert data from a standard velocity versus substrate concentration data plot (as in Figure 3.1), to a double reciprocal data plot as in Figure 3.2, you divide the calculated K_m value into 1 to obtain $1/K_m$. Also, when you plot your data on a double reciprocal data plot (Figure 3.2) you should notice that the intercept of the "straight line" intersects the horizontal (x) axis to the left of the y axis and therefore any calculated K_m value will show as a negative $1/K_m$ ($-1/K_m$). If you divide 0.5 mM into 1 you get a value of 2, and if you divide 0.05 into 1 you get a value of 20. Two is smaller than 20 and thus the line of the nonmodified mechanism will intersect the x axis toward the right and the line of the modified mechanism will intersect the x axis toward the left. How far apart they intersect will depend on how big the difference (how many orders of magnitude?) will be between the modified and nonmodified enzyme mechanisms. A similar point may be made for velocity values plotted on a double reciprocal data plot. Greater velocity values (i.e., 100) will show up as being lower on the y axis than a lesser velocity of, for example, 10. This point is one of the reasons why double reciprocal data plots are more difficult for the novice to interpret than the Eadie–Hofstee data plot illustrated in Chapter 9.

In our example above, we changed the relative values of the modified enzyme association rate constant relative to the nonmodified enzyme association rate constant. I think you can appreciate that you can obtain virtually the same outcomes if you change only the dissociation rate constant for the modified enzyme relative to the dissociation rate constant for the nonmodified enzyme. However, in this case you would need to make the dissociation rate constant for the modified enzyme *smaller* than the dissociation rate constant for the nonmodified enzyme in order to achieve a similar *activation* in the rate of conversion of substrate to product. There are two reasons why I am not going to get into this alternative approach to modifying the apparent K_m value to achieve an activation (or to achieve an inhibition) of the conversion of substrate to product. Firstly, I think you get the point and to go further would chance you getting frustrated. But secondly (and perhaps more importantly), I want to shift your attention to the role of the "affinity constant" (which I will define as K_{aff}), which under our quasi-equilibrium assumptions is quite analogous to a K_m value associated with the association and dissociation of the modifier Ma2 as it binds (associates with) to free enzyme E and dissociates from the modified enzyme–modifier complex (Ma2E) as illustrated in Equation 10.8.

When we think about a modifier binding to an enzyme in a reversible manner, we should immediately realize that depending on the relative values and relationships of the association and dissociation constants for modifier binding to an enzyme, we can create situations where most or very little of the enzyme can be present in a modified form. I have repeated Equation 10.7 below, but have renumbered it as Equation 10.9 because I wish now to focus your attention only on the association/dissociation of some modifier to free enzyme (the part of Equation 10.9 inside the dashed line area) and then isolated below as an independent set of association/dissociation events where modifier reversibly binds to free enzyme.

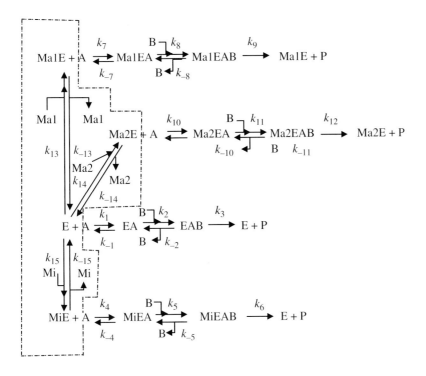

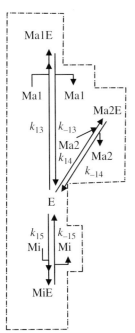

(10.9)

Equation 10.9 is provided in two parts so that you can see the whole equation from Equation 10.7, but I have delineated part of Equation 10.7 to show only the involvement of modifiers in shifting free enzyme E to a total of four forms of free enzyme that are capable of binding substrate A, first in what we have already described as an ordered sequential two-substrate one-product nonreversible conversion of substrate to product.

I am trying to impress you with how complicated the "commitment step" in the aromatic amino acid biosynthetic pathway is proving to be. As you get more involved in biochemistry, you will learn that the aromatic amino acid biosynthetic pathway is involved in the synthesis of more "products" than the three amino acids on which I have focused my discussion. However, I am not about teaching you biochemistry in this book, so I have chosen to restrict myself to an abbreviated description of the regulation of these three amino acids. Remember, what I have illustrated in Equation 10.7 represents the regulation of only one of the three D-arabinoheptulosonate-7-phosphate synthetase isozymes, and thus for just this single "commitment step" in aromatic amino acid biosynthesis, the enzyme mechanisms you would have to deal with would involve a total of three combinations of mechanisms as illustrated in Equations 10.7 AND 10.9.

Focusing on just the events shown in Equation 10.9, where a modifier reversibly binds to free enzyme E, you will note that there are a total of three equilibrium modifier-binding events likely to occur at any given time in intermediary metabolism associated with the synthesis of three amino acids, tyrosine, phenylalanine, and tryptophan. These three events are as illustrated below.

$$E + Mi \underset{k_{-15}}{\overset{k_{15}}{\longleftrightarrow}} MiE \qquad\qquad 1$$

$$E + Ma1 \underset{k_{-13}}{\overset{k_{13}}{\longleftrightarrow}} Ma1E \qquad\qquad 2$$

$$E + Ma2 \underset{k_{-14}}{\overset{k_{14}}{\longleftrightarrow}} Ma2E \qquad\qquad 3$$

What relative values are assigned (we assign, but nature does whatever it needs to do over time) to the rate constants in each of these three events will dictate how much free enzyme (E) is converted to MiE, Ma1E, and Ma2E and thus which pathway (mechanism in Equation 10.7) the two substrates will flow through. I hope you just thought of a really great question, which I will deal with later.

Just trying to imagine all possible options for relationships between the six rate constants above is enough to make you go blind. Not to mention, when you consider that these six rate constants are associated with only one of the three isozymes, and thus there are a total of 18 rate constants you would need to "imagine," it really becomes interesting. Therefore, try to think in a broad perspective. If, for example, you think a living cell will want to maximize the regulatory control over the metabolic activity of this enzyme, you might wish the following relationships to exist: $k_{15} > k_{-15}$, $k_{13} > k_{-13}$, and $k_{14} > k_{-14}$. If these relationships existed in cellular metabolism, there would be very little substrate produced by the nonmodified mechanism and the cell would possess maximal control over the synthesis of D-arabinoheptulosonate-7-phosphate. You might want to consider what other roles each of the three amino acids might have in cellular metabolism. Perhaps the cellular metabolism may need tryptophan more than it needs tyrosine and/or phenylalanine. In this particular case you may need to consider rate constants associated with the possible mechanisms associated with the isozyme that is feedback-inhibited by tryptophan more than you need to think about an opportunity to deal with a particular rate constant associated with the binding of tryptophan (to activate) to an isozyme that is feedback-inhibited by phenylalanine?

Not to forget that in an earlier discussion, I told you that unless otherwise specified, we would always assume that a given modifier of a given enzyme would be at "saturating" concentrations and present as an infinite quantity. This is one of those times when the modifier is probably not at "saturating" concentrations and also probably not present as some infinite quantity. Thus as the concentrations of these three amino acids (tryptophan, phenylalanine, and tyrosine) ebb and flow with the metabolic activity of the cell, the concentrations of modified and nonmodified enzyme will fluctuate to meet the needs of the cell; yet in all of this regulation, the concentrations of these three amino acids in cellular metabolism will remain remarkably constant even when any one of these three amino acids is added to the environment outside the cell—meaning the cell can also control how much of a metabolite is transported across the cell membrane using comparable regulatory means.

Do you still have that question in your head? I hope that question was something along the line of: *"If all three isozymes are making the same substrate through any one of 12 possible enzyme mechanisms and that one substrate (D-arabinoheptulosonate-7-phosphate) is ultimately metabolized to all three amino acids, why does it matter which enzyme mechanism, modified or not modified, is involved in the synthesis of this particular metabolite?"* Was that your question?

If that was your question, you are understanding the objectives of this book. So what is the answer? The answer leads us to the next level of discussion of regulation of enzymes by intermediates (or "end product" for the immediate purpose of this section of our discussion) in the aromatic amino acid biosynthetic pathway. We need to look further along in the aromatic amino acid biosynthetic pathway for a part of the answer. Let's look again at Figure 4.6, repeated here again as Figure 10.4. I chose Figure 4.6 only for convenience in that we have been covering the use of phenylalanine as a modifier in prior discussions, but I could just as easily used either of the other two figures (Figures 4.5 and 4.7) in Chapter 4. If you look carefully at Figure 10.4 (= Figure 4.6) you will observe that P_2 (phenylalanine in Figures 4.5, 4.6, and 4.7) exerts a modifier effect on enzymes e, f, h, and i in addition to enzyme b (which we have been discussing as a "commitment step enzyme." Phenylalanine exerts a "positive" (activation) effect on enzyme e such that it would tend to increase the production of tryptophan from earlier metabolites in this pathway, and on enzyme i such that it would tend to "increase" the production of tyrosine from earlier metabolites. However, phenylalanine exerts a "negative" (inhibition) effect on enzymes f and h.

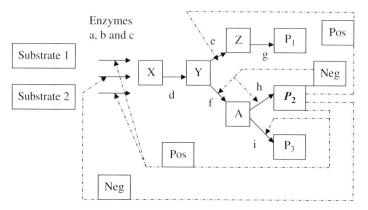

Figure 10.4 Repeated Figure 4.6. Diagram of the aromatic amino acid biosynthetic pathway showing the enzymatic steps in the pathway at which phenylalanine (P_2) modifies enzymatic activity via positive and negative regulatory events.

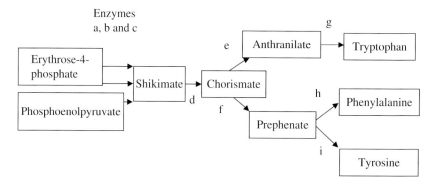

Figure 10.5 Repeated Figure 4.4. Pathway as shown in Figure 4.3, but with specific molecules found in this aromatic amino acid biosynthetic pathway substituted for the symbols used in Figure 4.3. Prephenate is actually metabolized to either phenylpyruvate prior to formation of phenylalanine or to parahydroxyphenylpyruvate prior to formation of tyrosine, but these intermediates were left out of this figure in order to emphasize the enzyme steps relevant (important) to regulation of this metabolic pathway and to aid in comprehension.

I have also again reproduced Figure 4.4 (as Figure 10.5) immediately below Figure 10.4 so that you do not have to look back through the book to determine what enzymes e, f, h, and i are converting in the way of substrate to product. But I again remind you that I have shortened this metabolic pathway to facilitate the discussion of enzyme regulation rather than accurately describing specific enzyme steps in this pathway. This book is supposed to be more about enzymology than about biochemistry.

I want you to focus your attention on enzymes f, h, and i in Figure 10.4 because this step is important in illustrating how regulation of these three enzymes by phenylalanine focuses on shifting intermediates in this pathway toward tyrosine. Also, by acting in a negative matter on enzyme f and on a positive manner on enzyme e, any reduction in the production of intermediate A by phenylalanine will tend to shift more metabolite Y toward the production of tryptophan through a positive effect on enzyme e, but also shift more of metabolite A toward tyrosine through a positive effect on enzyme i. Remember that I have taken liberties with this pathway, and prephenate (metabolite A in Figure 10.4) is actually metabolized to either phenylpyruvate prior to formation of phenylalanine or to parahydroxyphenylpyruvate prior to formation of tyrosine, but these intermediates were left out of this figure in order to emphasize the enzyme steps relevant (important) to regulation of this metabolic pathway and to aid in comprehension.

If you feel that this is all too complicated, remember that you have to "overlay" Figures 4.5, 4.6, and 4.7 in order to understand that many (most) of these enzymes are similar to the DAHP synthetase isozymes (in the "commitment step") in being both positively and negatively regulated (activated and inhibited) by each of these three amino acids. Thus, the earlier equation(s), for example Equation 10.7, could also be used to describe possible regulatory mechanisms for enzymes later in the aromatic amino acid biosynthetic pathway—for example enzyme f, but you have to be careful in determining which modifier is exerting a positive (activation) or negative (inhibition) on which enzyme in the overall pathway.

How does the above pertain to a Roman water distribution system? You can think about just one of the pipes coming off the main distribution pipe in Figure 4.1 (let's pick pipe A). Think about a water pipe carrying water to a single "house," but that house has three fountains for the water to run into. It may not be important that all three fountains run at the same time, but it may be important that you can turn off the fountain closest to your bed at night, while avoiding extra flow through the other two fountains at night—because you have to pay for the water that runs through your house and into "public drains." (Are drains another enzyme opportunity? Where do phenylalanine, tyrosine, and tryptophan go once they are synthetized?) Remember, the water coming into your fountains provides you with drinking and cooking water, but the water you do not use probably flushes the sewage from your toilets into the local river or ocean.

You could install shut-off or metering valves to regulate the water entering your household water distribution system, and also install shut-off or metering valves at each fountain—if you had shut-off or metering valves in those days. If you didn't, you could regulate water flow within your house by doing two things with your fountains. You could use differently sized pipes feeding water to each fountain (regulate "V_{max}") and/or you could use differing heights for the outflow of water into the fountains (regulate "K_m"). If you wanted to regulate the flow of water through a given fountain, you could add or remove a specific length of pipe to/from the nonmodified pipe and/or you could add or remove a specific length of pipe that was reduced in inside diameter to/from the nonmodified pipe feeding that specific fountain. You would never wish to completely stop the flow of water through the fountain, because then "other metabolites" might pile up (pardon the expression) and that might be unsanitary.

Intermediary metabolism is a complicated network of enzymes, enzyme synthesis and degradation, metabolites, and metabolite concentrations, and requires a constant flow of these metabolites into, through, and out of the cell in order for a cell to thrive and reproduce. If you strive to memorize the role(s) and functions of specific enzymes rather than learning to accept the complexity of an enzyme in being highly modified by metabolites you will never get the pleasure of trying to figure out the details of how that enzyme can possibly be modified to achieve what needs to be accomplished.

11

Rate Equation Derivation by the King–Altman Method

Two Substrates and Two Products

> *The ready availability of information is useless if you don't know what to do with it or how to use it.*
>
> L. Wolfinbarger, Jr, 1998

I'm not sure whether you have noticed the italicized quotations at the beginning of each chapter, but the quote at the top of this chapter is one I frequently put on the top of the "notes" I provided to my students that contained the important points I intended to make in my lecture for that day. The idea originated in my biochemistry lectures where my students frequently challenged my require-ment that they learn the structures of the 20 common amino acids found in proteins. Many students objected to learning these structures, advising me that if they needed to know the structure of a given amino acid (or for that matter the structure of virtually any organic compound), they could simply look it up via the "internet" or in their textbook. My standard response was that knowing the structures was not the important part of the exercise. The important part of the exercise lay in the "imprinting" of the differences between the amino acids in their minds. That way when someone asked them why a particular pH gradient was necessary for eluting such compounds from a cation-exchange resin and in which order the amino acids might elute (such as was used in the old days of amino acid analysis—those were the days when you had to walk to school through the snow and it was uphill both directions), they could appear intelli-gent (meaning smarter than the questioner) by not having to say: "I'm not sure, let me look that up."

So it is with enzyme mechanisms and rate equation derivations. The details of some mechanism (or rate equation derivation) is not the important part of the exercise. The important part of the exercise is being able to think your way through the possibilities and being able to explain why you chose some combination(s) of rate constants and/or mechanisms for modification of some enzymatic reaction and "appear intelligent." The "being intelligent" part of this lies in knowing what to do with the information you have before you.

Enzyme Regulation in Metabolic Pathways, First Edition. Lloyd Wolfinbarger.
© 2017 John Wiley & Sons, Inc. Published 2017 by John Wiley & Sons, Inc.

We delay getting into the interesting stuff. Let's get on with the King–Altman method for deriving a rate equation for a two-substrate and two-product enzyme reaction. I'm going to keep this simple so basically we will be looking at the following:

$$A + B \leftrightarrow Q + P$$

Of course very little in this chapter will be "simple," but I like to think that once you get into the chapter, you will already be familiar with the symbols and concepts, so hopefully you will be able to draw on what you previously learned to make this chapter "simple."

First of all, let me take you back to a previous equation you have already seen. The equation was Equation 10.3, which I will repeat here as Equation 11.1.

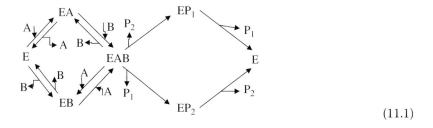

$$(11.1)$$

However, let's change this equation to that provided below as Equation 11.2:

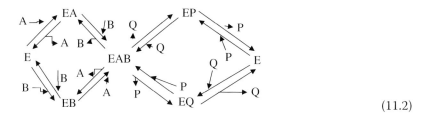

$$(11.2)$$

The only differences between Equation 11.1 (10.3) and Equation 11.2 lie in making the formation of the two products reversible events and in assigning a letter Q in place of "P_2" and a letter P in place of "P_1." This enzyme mechanism implies that the sequence of binding of the two substrates is random, that the sequence of release of the two products is also random, and that all the steps in the overall enzyme mechanism are reversible. I slipped Equation 11.2 in at this point in the chapter to illustrate an extremely complicated two-substrate and two-product enzyme mechanism and to let you know that we won't be deriving a rate equation for this mechanism as a beginning exercise. I propose that we take on a far less complicated two-substrate and two-product enzyme mechanism for your first adventure.

Let's start with a two-substrate and two-product enzyme mechanism where the additions of substrates are ordered (meaning one substrate will combine with free enzyme before that enzyme–substrate complex will bind with the second substrate). Let's also start with the release of products as ordered (meaning one product will be released from the enzyme–substrate(s) complex before the second product will be released. Let's, however, keep all steps in the overall enzyme mechanism reversible. Let's create Equation 11.3:

$$E + A \rightleftarrows EA + B \rightleftarrows (EAB/EPQ) \rightleftarrows EQ + P \rightleftarrows Q + E \qquad (11.3)$$

I have left off rate constants in Equation 11.3 while I make the following points:

1) We will be operating at "saturating" concentrations of substrates and products (meaning we will not be deriving a rate equation where no product has been formed), but rather where the enzyme mechanism is at "equilibrium" in the interconversion of substrates to products.
2) The amounts of substrates and products will be "inexhaustible."
3) We will not be assigning relative values to kinetic constants (which we will assign in a later rendition of Equation 11.3).
4) For now, whether we are operating under quasi-equilibrium assumptions or not is not important.

Let's add kinetic constants (Equation 11.4) and make a few additional comments:

$$E + A \underset{k_{-1}}{\overset{k_1}{\rightleftarrows}} EA + B \underset{k_{-2}}{\overset{k_2}{\rightleftarrows}} (EAB/EPQ) \underset{k_{-3}}{\overset{k_3}{\rightleftarrows}} EQ + P \underset{k_{-4}}{\overset{k_4}{\rightleftarrows}} Q + E \qquad (11.4)$$

The comments here are not complicated. I simply wish to emphasize that the "EAB/EPQ" is written in Equation 11.4 only to emphasize that we will operate as though the actual conversion of the two substrate molecules to the two product molecules is not relevant to the rate equation derivation (the transition is "instantaneous" and will hereafter simply be designed as "EAB" in future mechanisms).

I would also like you to note that many people write the final product release as "E + Q" rather than the "Q + E" as written in Equation 11.4. Writing this mechanism as I have in Equation 11.4 was actually one of those "Aha!" moments for me where the mechanism (as with all enzyme mechanisms) begins and ends with "free enzyme." I had always learned and wrote my mechanism using the "E + Q" because that was how I began a mechanism ("E + A"). Making the simple change of replacing the "E + Q" with the "Q + E" enabled me to finally grasp why the King–Altman method used a different approach to writing Equation 11.4. King and Altman wrote Equation 11.4 in the form of a closed "box" rather than in the linear fashion you have seen throughout this book. The linear way of writing such mechanisms is great, but writing such a mechanism as a "box" finalizes the understanding that all enzyme mechanisms begin and end with "free enzyme."

Let's redraft Equation 11.4 in a "box" format, which we shall call Equation 11.5:

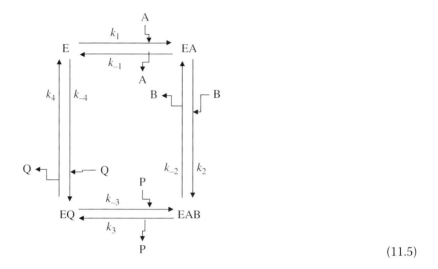

$$(11.5)$$

If you go step by step, you will see that Equation 11.4 is virtually identical to Equation 11.5 with the exception that the former equation is in a linear format and the latter equation is in a box (or circular) format. The only minor changes between the two formats lie in the box format showing where each substrate and each product adds to an enzyme/enzyme complex or is released from an enzyme/enzyme complex.

Now that you are comfortable with what I have done, I want to change Equation 11.5 to Equation 11.6 so I can better illustrate the mechanism to show where substrate and product concentrations exert an influence on any subsequent rate equation derivation. In short, the concentration of substrate "A" will influence the rate at which the "E" (free enzyme) will change to the "EA" (or enzyme–substrate "EA") complex and thus some "velocity" or "rate" will be satisfied by velocity = $[A]k_1$. In order for me to mimic the way this "box" diagram is normally presented in descriptions of the King–Altman method, I will change the "velocity = $[A]k_1$" to "velocity = ak_1," but you are free to represent this event however best suits your needs. Since the concentration of product (the "product" in the sense of the reverse reaction associated with k_{-1}) A will not influence the rate of conversion of the enzyme–substrate EA back to free enzyme E and free substrate A, there is no velocity measurement associated with k_{-1} to be influenced by the concentration of A (i.e., $[A]$).

Now, I am going to write three additional paragraphs that you will find to be virtually identical to the paragraph dealing with substrate A binding to free enzyme, but each separate paragraph will deal with the binding of substrate B, "substrate P" (note the quotation marks surrounding "substrate P"), and "substrate Q" (again note the quotation marks surrounding "substrate Q"). You will need to keep in mind that for the forward reactions, A and B represent the substrates, but for the backward reactions, "P" and "Q" represent the substrates.

If you understand what I did with respect to the concentration of substrate [A] influencing the velocity of the conversion of A plus E to the EA complex and can apply this same understanding to the following three paragraphs, you might want to skip the next three paragraphs.

Again, under normal circumstances we would think of compounds P and Q as "products," but in the "reverse direction" of the enzymatic conversion of substrates A and B to products P and Q, products P and Q become the substrates and substrates A and B become the products.

Personally, I would not skip these next three paragraphs. I am writing this material and I have to continually check behind myself to make sure that I put the correct values in places where I am changing values.

Skip? I want to change Equation 11.5 to Equation 11.6 such that I can better illustrate the mechanism to show where substrate and product concentrations exert an influence on any subsequent rate equation derivation. In short, the concentration of substrate B will influence the rate at which the EA (enzyme–substrate A) complex will change to the EAB (or enzyme–substrate AB) complex and thus some "velocity" or "rate" will be satisfied by velocity $= [B]k_2$. In order for me to mimic the way this "box" diagram is normally presented in descriptions of the King–Altman method, I will change the "velocity $= [B]k_2$" to "velocity $= bk_2$," but again you are free to represent this event however best suits your needs. Since the concentration of B, that is, product [B], will not influence the rate of conversion of the enzyme–substrate AB complex to the enzyme–product EA complex and free substrate B, there is no velocity measurement associated with k_{-2} to be influenced by the concentration of B (i.e., [B]).

Skip? I want to change Equation 11.5 to Equation 11.6 such that I can better illustrate the mechanism to show where substrate and product concentrations exert an influence on any subsequent rate equation derivation. In short, the concentration of substrate P will influence the rate at which the EQ (enzyme–substrate Q) complex will change to the EAB (or enzyme–substrate AB) complex and thus some "velocity" or "rate" will be satisfied by velocity $= [P]k_{-3}$. In order for me to mimic the way this "box" diagram is normally presented in descriptions of the King–Altman method, I will change the "velocity $= [P]k_{-3}$" to "velocity $= pk_{-3}$," but you are again free to represent this event however best suits your needs. Since the concentration of substrate P (i.e., [P]) will not influence the rate of conversion of the enzyme–substrate AB complex to the enzyme EQ complex and free product P, there is no velocity measurement associated with k_3 to be influenced by the concentration of P (i.e., [P]).

Skip? I want to change Equation 11.5 to Equation 11.6 such that I can better illustrate the mechanism to show where substrate and product concentrations exert an influence on any subsequent rate equation derivation. In short, the concentration of substrate Q will influence the rate at which the free enzyme (E) will change to the EQ complex by velocity $= [Q]k_{-4}$. In order for me to mimic the way this "box" diagram is normally presented in descriptions of the King–Altman method, I will change the "velocity $= [Q]k_{-4}$" to "velocity $= qk_{-4}$," but you are again free to represent this event however best suits your needs. Since the concentration of substrate Q (i.e., [Q]) will not influence the rate of conversion of the enzyme product EQ complex to the free enzyme E and free product Q, there is no velocity measurement associated with k_4 to be influenced by the concentration of Q (i.e., [Q]).

We are now ready to change Equation 11.5 to Equation 11.6. You will note that Equation 11.6 shows only the "addition" of substrate molecules in association with the appropriate kinetic constant. Remember, that for the "forward" reaction, that is, conversion of compounds A and B to compounds P and Q, A and B are denoted as substrates and P and Q are denoted as products. For the "backward" reaction, that is, conversion of compounds P and Q to compounds A and B, P and Q are denoted as substrates and A and B are denoted as products. I trust you picked these points out in the above four paragraphs?

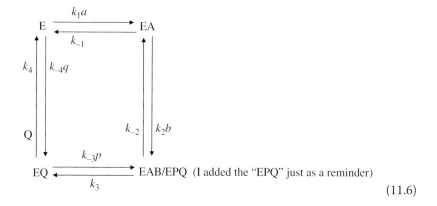

$$(11.6)$$

We have a much more simple representation of a two-substrate and two-product reversible enzyme mechanism in Equation 11.6 than we had in Equation 11.5, and with this simpler representation I can move onto an explanation of the King–Altman method for deriving a rate equation for a reversible two-substrate and two-product enzyme reaction.

One of the first things you should note in Equation 11.6 is there are only four enzyme species: 1) species E; 2) species EA; 3) species EAB; and 4) species EQ. There are no species EB or EP since B and P do not bind to free enzyme in this ordered mechanism. The second thing you should note in Equation 11.6 is that the reactions between the four enzyme species are readily evident. All of the reactions must be treated as first-order reactions where the second-order rate constants, for example k_1, k_2, k_{-3}, and k_{-4}, are replaced by what I will call pseudo-first-order rate constants, that is, k_1a, k_2b, $k_{-3}p$, and $k_{-4}q$, by including the concentrations of compounds "a", "b", "p", and "q" in describing the reactions. I stipulate "pseudo-first-order rate constants" in this description in that while we may claim "saturating" concentrations of these substrates in the enzyme reaction, as you will remember from earlier chapters, one can approach a maximal velocity at infinitely high concentrations of substrates, but one never achieves true V_{max} and thus one never achieves a true first-order event, only a pseudo-first-order event. For purposes of this book, I will equate pseudo-first-order as being first order since it would be difficult to demonstrate a mathematical difference in some arrived at velocity value for a rate equation derivation.

You can now show Equation 11.6 as a simple "box" illustration—a basic "square" (Equation 11.7)—the "Master Pattern."

(11.7)

Equation 11.7 does not look like an equation, but as you will see, the four sides of the box represent the four sides of Equation 11.6 and thus each of the four sides of the box in Equation 11.7 can be used to represent one of the four enzyme species and the rate constants associated with a given side (see the "Master Pattern"). We go "simple" before we go back to "complicated".

The method of King and Altman stipulates that you find every pattern that 1) consists only of lines from the Master Pattern, 2) connects every enzyme species, and 3) contains no closed loops. Thus, each pattern will contain one line fewer than the number of enzyme species. Fortunately for us, there are only four such patterns (Equation 11.8).

You can draw these four patterns in any order you wish. I have drawn them in the order given because the "up," "down," and two facing each other seems to work best for me.

(11.8)

For each enzyme species, and each pattern, the product of the rate constants in the pattern leading to that species can be written down where the combinations of three "arrows" (as illustrated in Equation 11.6) all point to only one of the four enzyme species. I'll give you a hint by telling you there are four such combinations of arrows for each of the above four patterns.

Patterns Ending on "E"

As an example, if you pick the enzyme species E for all three arrows to point to, you will find that one combination of arrows from each of the four patterns illustrated in Equation 11.8. I will illustrate these four patterns showing only the arrows, in Equation 11.9:

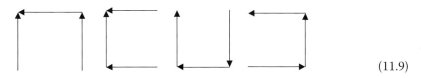

(11.9)

The kinetic constants associated with each of the three "arrows" in each of the four patterns correspond to the following (going from left to right): 1) $k_4\,k_{-1}\,k_{-2}$; 2) $k_4\,k_{-1}\,k_3$; 3) $k_4\,k_3\,k_2$, and 4) $k_{-3}\,k_{-2}\,k_{-1}$. These combinations of kinetic constants are shown associated with their respective arrows in Equation 11.10:

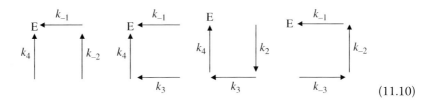

$$(11.10)$$

We come to a point of teaching style. Do we continue with all combinations of kinetic constants ending on E and describe which substrates will be associated with which kinetic constant—for example, we know that Q will eventually be multiplied by kinetic constant k_{-4} (but not k_4)—and we will eventually need to begin stringing together a series of combinations of kinetic constants with substrate multipliers. However, I think it best if we continue doing what you have just worked on, figure out the other 12 combinations of "arrows" ending on EA, EAB, and EQ and then return to the next step of stringing these series of combinations of kinetic constants with substrate multipliers where we have more to work with. I mostly just wanted to pause here and let you know where we were going after this and to give you time to think about (determine) whether or not you can figure out the other 12 combinations of "arrows" on your own.

Patterns Ending on "EA"

Did you figure out the other 12 combinations? If you have, we should check your "figuring," and if not we should go ahead and figure them out at this time. Let's go on to the enzyme species "EA." There should be four such patterns, all ending on EA, and there should be one combination of arrows from each of the four original patterns shown in Equation 11.8. The four patterns ending on EA are as shown in Equation 11.11:

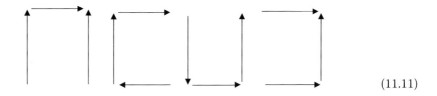

$$(11.11)$$

The kinetic constants associated with each of the three "arrows" in each of the four patterns ending on EA correspond to the following (going from left to right): 1) $k_1\,k_4\,k_2$; 2) $k_1\,k_4\,k_3$; 3) $k_{-4}\,k_{-3}\,k_{-2}$; and 4) $k_1\,k_{-3}\,k_{-2}$. These combinations of kinetic constants are shown with their respective "arrows" ending on EA in Equation 11.12:

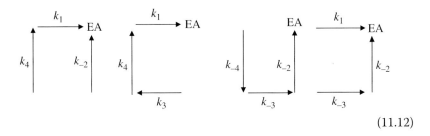

$$(11.12)$$

Just as a quick reference, have you realized that of all the kinetic constants shown in Equation 11.12, we will eventually link to substrate A with the k_1 kinetic constant, substrate Q with the k_{-4} kinetic constant, and substrate P with the k_{-3} kinetic constant? While I wish to wait until later to get further into these issues, it is important that you begin to think about these issues while we work out the combinations of "arrows" and kinetic constants in the four patterns for each enzyme species.

Patterns Ending on "EAB"

Let's go onto the enzyme species "EAB." There should be four such patterns, all ending on EAB, and there should be one combination of arrows from each of the four original patterns shown in Equation 11.8. The four patterns ending on EAB are as shown in Equation 11.13:

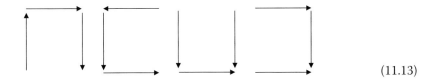

$$(11.13)$$

The kinetic constants associated with each of the three "arrows" in each of the four patterns ending on EA correspond to the following (going from left to right): 1) $k_1 k_4 k_2$; 2) $k_{-1} k_{-4} k_{-3}$; 3) $k_{-4} k_{-3} k_2$; and 4) $k_1 k_{-3} k_2$. These combinations of kinetic constants are shown with their respective "arrows" ending on EA in Equation 11.14:

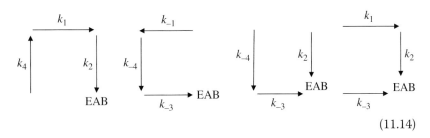

$$(11.14)$$

Again, just as a quick reference, have you realized that of all the kinetic constants shown in Equation 11.14, we will eventually link substrate A with the k_1 kinetic constant, substrate Q with the k_{-4} kinetic constant, substrate P with the k_{-3} kinetic constant, and substrate B with the k_2 kinetic constant.

Patterns Ending on "EQ"

Let's go onto the final enzyme species "EQ." There should be four such patterns, all ending on EQ, and there should be one combination of arrows from each of the four original patterns shown in Equation 11.8. The four patterns ending on EQ are as shown in Equation 11.15:

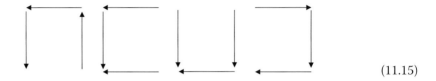

$$(11.15)$$

The kinetic constants associated with each of the three "arrows" in each of the four patterns ending on EQ correspond to the following (going from left to right): 1) $k_{-1}\ k_{-4}\ k_{-2}$; 2) $k_{-1}\ k_{-4}\ k_3$; 3) $k_{-4}\ k_3\ k_2$; and 4) $k_1\ k_3\ k_2$. These combinations of kinetic constants are shown with their respective "arrows" ending on EQ in Equation 11.16:

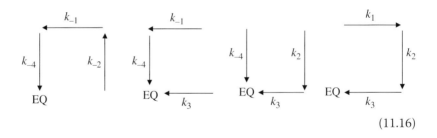

$$(11.16)$$

Again, just as a quick reference, have you realized that of all the kinetic constants shown in Equation 11.16, we will eventually link substrate A with the k_1 kinetic constant, substrate Q with the k_{-4} kinetic constant, and substrate B with the k_2 kinetic constant?

Have you noticed yet, that in each of the patterns of arrows in each of the 16 patterns of three arrows per pattern, all arrows begin on the enzyme species that are not the enzyme species to which the arrows point? Remember that there are four combinations of three arrows pointing to each of the four enzyme species, and all three arrows pointing to each of the four enzyme species originate from enzyme species other than the enzyme species to which they point. Then you will have a basis for remembering the fundamental rules of the King–Altman method.

Finally (?), let's turn to setting up a rate equation that will define an ordered two-substrate and two-product enzyme reaction. We have all the fundamental combinations of kinetic constants and substrate multipliers that we need, we just have to figure out how to organize them in a rate equation. As you will see, the fraction of each enzyme species in the steady-state mixture will be some combination of the sum of its four kinetic constant combinations plus the substrate multiplier(s) divided by the sum of all 16 kinetic constant combinations plus the substrate multipliers. But, it is not quite that simple and we get ahead of ourselves. We need firstly to organize the sums of the kinetic constants (plus any substrate multiplier(s)) for each of the four enzyme species. If you go back to the combinations of kinetic constants for each of the enzyme species we developed earlier, all you need to do is add the four sets of three kinetic constants (plus any substrate multiplier and set them equal to the concentration of that respective enzyme species (where that concentration is determined by the fraction of that enzyme species in the total concentration of all enzyme species) and divide that concentration by the total concentration of enzyme species (which we will designate as "[E_t]").

I appreciate that there were a "lot" of concepts to think about in the paragraph above, but if you go back to the section "Patterns Ending on "E"" and look at Equation 11.9 (plus the paragraph just above Equation 11.9) you will see the four sets of three kinetic constants associated with each of the four combinations of "arrows". You will also find a description in the text after Equation 11.9 that stipulates which substrate concentrations are associated with which kinetic constant. If you add these four sets of kinetic constants (where the substrate multiplier is included), you will gain the following equation (Equation 11.17):

$$[E]/[E_t] = (k_{-1}k_{-2}k_4 + k_{-1}k_3k_4 + k_2k_3k_4b + k_{-1}k_{-2}k_{-3}p)/\Sigma \qquad (11.17)$$

You probably noticed that I slipped in a divider designated at Σ which I have not used before. We have not established what this Σ symbol means yet, but basically it will be the sum of all the "proved" kinetic constant combinations (associated with their respective substrate multiplier(s)) for each of the four enzyme species. Wait for it. It will become more clear when we have finished constructing the four equations defining the fractions each enzyme species contributes to the overall rate equation derivation.

Let's do the same kind of consolidation of the "Patterns Ending on "EA"" that we did for the "Patterns Ending on "E"". You will find the patterns for "EA" in Equation 11-11 and in the paragraph just before that equation. You will also find the substrate multipliers and which kinetic constant they will be associated with in the paragraph following that equation.

If you add these four sets of kinetic constants (where the substrate multipliers are included) you will obtain Equation 11.18:

$$[EA]/[E_t] = (k_1k_{-2}k_4a + k_1k_3k_4a + k_{-2}k_{-3}k_{-4}pq + k_1k_{-2}k_{-3}ap)/\Sigma \quad (11.18)$$

Let's do the same kind of consolidation of the "Patterns Ending on "EAB"" that we did for the "Patterns Ending on "E" and the "Patterns Ending on "EA."" You will find the patterns for EAB in Equation 11.14 and in the paragraph just before that equation. You will also find the substrate multipliers and which kinetic constant they will be associated with in the paragraph following that equation if you can't figure them out yourself.

If you add these four sets of kinetic constants (where the substrate multiplier is included), you will get the following equation (Equation 11.19):

$$[EAB]/[E_t] = (k_1 k_2 k_4 ab + k_{-1} k_{-3} k_{-4} pq + k_2 k_{-3} k_{-4} bpq + k_1 k_2 k_{-3} abp)/\Sigma$$
(11.19)

Let's do the same kind of consolidation of the "Patterns Ending on "EQ"" that we did for the "Patterns Ending on "E," the "Patterns Ending on "EA," and the "Patterns Ending on "EAB."" You will find the patterns for EQ in Equation 11.16 and in the paragraph just before that equation. You will also find the substrate multipliers and which kinetic constant they will be associated with in the paragraph following that equation if you can't figure them out yourself.

If you add these four sets of kinetic constants for patterns ending on EQ (where the substrate multiplier is included), you will get the following equation (Equation 11.20):

$$[EQ]/[E_t] = (k_{-1} k_{-2} k_{-4} q + k_{-1} k_3 k_{-4} q + k_2 k_3 k_{-4} bq + k_1 k_2 k_3 ab)/\Sigma \qquad (11.20)$$

From here on, you might think things are really going to get complicated, but in actuality, they don't. It just becomes tedious keeping track of kinetic constants and substrate multipliers as you try to develop some velocity (or rate) equation dealing with the change in concentration(s) of the individual substrates or products from this particular enzyme mechanism. For example, in Equation 11.4 (reproduced below as Equation 11.4 Repeated), you will notice that products P and Q can only be formed by the consumption of substrates A and B. Conversely, products A and B can only be formed by the consumption of substrates P and Q. Thus there is a one-to-one linkage between substrates and products whether you are measuring some "forward" reaction or some "backward" reaction (to use laboratory jargon). You can therefore measure the disappearance of one or both substrates or the appearance of one or both of the products to determine a velocity of the enzymatic reaction.

$$E + A \rightleftharpoons EA + B \rightleftharpoons (EAB/EPQ) \rightleftharpoons EQ + P \rightleftharpoons Q + E$$
(11.3 repeated)

As an example, let's measure the appearance (formation) of "product P." We can define the velocity (or rate) of formation of P as the change in the concentration of P (d[P]) as a function of change in time (dt) (see Figure 9.2 for data illustrating this point). Further, the rate of the enzymatic reaction leading to the formation of P is the sum of the rates of the steps that generate P minus the sum of the rates of the steps that consume P. In our case, the step that generates P is the EAB conversion to EQ through kinetic constant k_3, and the step that consumes P is the EQ conversion to EAB through kinetic constant k_{-3}. However, as you remember, the kinetic constant k_{-3} carries a multiplier "p" (which corresponds to the concentration of P, i.e., [P]). Thus we end up with Equation 11.21:

$$\text{Velocity}(v) = d[P]/dt = k_3[EAB] - k_{-3}[EQ]p \tag{11.21}$$

However, EAB and EQ would not be present except for the other steps in Equation 11.4; therefore, to satisfy the overall enzyme mechanism illustrated in Equation 11.4, we must factor in additional kinetic constants involved in the formation of P and in the consumption of P by the total enzyme species in the enzymatic reaction. Thus we get an expanded velocity equation in the form of Equation 11.22:

$$\text{Velocity}(v) = d[P]/dt = E_t\left(k_1 k_2 k_3 k_4 ab - k_{-1}k_{-2}k_{-3}k_{-4}pq\right)/\Sigma \tag{11.22}$$

I suspect you can look at Equation 11.4 and easily see that the kinetic constants k_1, k_2, and k_3 are obviously involved in the formation of product P, but why k_4? Like yourselves I was originally confused, but finally understood that unless EQ went on to release Q and free enzyme E, we would soon run out of free enzyme and no longer be able to produce product P, hence the role of kinetic constant k_4. Similarly for the role of kinetic constant k_{-4} in the consumption of P. You cannot consume P in the reverse mechanism of the enzyme mechanism illustrated in Equation 11.4 if you do not have the EQ enzyme species to bind with P resulting in the formation of A and B. As I described earlier, there is a one-to-one linkage of substrates and products in the enzyme mechanisms illustrated in Equation 11.4. If you wish to measure the velocity of the conversion of substrate A or B to products P or Q, you can also measure the consumption of either A or B in the same manner in which you measured the appearance of P (or Q). Simply substitute d[A]/dt, d[B]/dt, or d[Q]/dt into Equation 11.22 and you will get the same velocity values as with d[P]/dt.

Finding "Σ"

Earlier I described "Σ" as the sum of "proved" combinations of kinetic constants (multiplied by appropriate substrate concentrations) from the "possible" combinations of kinetic constants associated with the enzyme reactions of each of the four enzyme species. However, I also told you this is not a simple thing to calculate, and thus I put you off until now. Equation 11.22 provides you with the numerator in the rate equation and because calculating the "proved" denominator (or Σ) is beyond the scope of this book I am going to provide you with the denominator value and attempt to show you some of the logic used in its calculation.

In that I am going to be changing how I describe the combinations of kinetic constants (and substrate multipliers) in order to reduce the complexity of the equations I provide to you, I will be returning several times to Equation 11.4 in order to keep you focused. However, I will also be manipulating Equation 11.4 to illustrate specific points as we develop the combinations of kinetic constants (and substrate multipliers). So try to keep focused on the simple illustration shown in Equation 11.4 of an ordered two-substrate and two-product enzyme mechanism that is reversible. I will begin by repeating Equation 11.4.

Reviewing Equation 11.4, you can see that the forward reactions (those where substrates A and B are converted to products P and Q) are as shown in Equation 11.23, and the backward reactions (those where the substrates P and Q are converted to products A and B are as shown in Equation 11.24.

$$E + A \underset{k_{-1}}{\overset{k_1}{\rightleftarrows}} EA + B \underset{k_{-2}}{\overset{k_2}{\rightleftarrows}} (EAB/EPQ) \underset{k_{-3}}{\overset{k_3}{\rightleftarrows}} EQ + P \underset{k_{-4}}{\overset{k_4}{\rightleftarrows}} Q + E$$

(11.4 repeated)

$$E + A \xrightarrow{k_1} EA + B \xrightarrow{k_2} (EAB/EPQ) \xrightarrow{k_3} EQ + P \xrightarrow{k_4} Q + E \qquad (11.23)$$

$$E + A \xleftarrow{k_{-1}} EA + B \xleftarrow{k_{-2}} (EAB/EPQ) \xleftarrow{k_{-3}} EQ + P \xleftarrow{k_{-4}} Q + E \qquad (11.24)$$

These latter two equations provide us with another way to express the velocity of the overall reaction of this two-substrate and two-product enzyme mechanism. The velocity of the forward reaction will include the sum of the four kinetic constants k_1, k_2, k_3, and k_4 multiplied by $[E_t]$ and the concentrations of substrates A and B, previously denoted as [A] and [B], but also simply as a and b.

The velocity of the backward reaction of this two-substrate and two-product enzyme mechanism will include the sum of the four kinetic constants k_{-1}, k_{-2}, k_{-3}, and k_{-4} multiplied by $[E_t]$ and the concentrations of the substrates P and Q, previously denoted as [P] and [Q], but also simply as p and q.

At some equilibrium, or near-equilibrium state, the velocity of the enzymatic reaction will be represented as the forward reaction minus the backward reaction, and we can write this as shown in Equation 11.25:

$$\text{Velocity}(v) = [\text{Et}]((k_1 + k_2 + k_3 + k_4)ab - (k_{-1} + k_{-2} + k_{-3} + k_{-4})pq) / \Sigma$$

(11.25)

It is getting increasingly difficult for me to write these equations and for you to read them, and thus if we can agree that $c1 = (k_1 + k_2 + k_3 + k_4)$ and that $c2 = (k_{-1} + k_{-2} + k_{-3} + k_{-4})$ we can rewrite Equation 11.25 as Equation 11.26:

$$\text{Velocity}(v) = [E_t](c1ab - c2pq) / \Sigma \qquad (11.26)$$

where $c1 = (k_1 + k_2 + k_3 + k_4)$ and $c2 = (k_{-1} + k_{-2} + k_{-3} + k_{-4})$

What I have done in Equation 11.26 is express the equation in coefficient form, making the equations simpler and easier to write and read. It also provides me with the ability to draft out the full velocity equation of the two-substrate and two-product ordered enzyme reaction as Equation 11.27:

$$v = \frac{[E_t]\left(c1ab - c2pq\right)}{\begin{array}{l} c3 + c4a + c5b + c6p + c7q + c8ab + c9ap + c10bq \\ k + c11pq + c12abp + c13bpq \end{array}} \tag{11.27}$$

where (remember c1 and c2 were provided earlier in Equation 11.26)

$$c3 = k_{-1}(k_{-2} + k_3)k_4$$
$$c4 = k_1(k_{-2} + k_3)k_4$$
$$c5 = k_2 k_3 k_4$$
$$c6 = k_{-1}k_{-2}k_{-3}$$
$$c7 = k_{-1}(k_{-2} + k_3)k_{-4}$$
$$c8 = k_1 k_2(k_3 + k_4)$$
$$c9 = k_1 k_{-2}k_{-3}$$
$$c10 = k_2 k_3 k_{-4}$$
$$c11 = (k_{-1} + k_{-2})k_{-3}k_{-4}$$
$$c12 = k_1 k_2 k_{-3}$$
$$c13 = k_2 k_{-3}k_{-4}$$

Where did all the terms in the denominator in Equation 11.27 come from? As I indicated earlier, determination of the approximate (depends on details of the enzyme mechanism, which I have not covered) 96 patterns required by the King–Altman analysis requires the elimination of patterns not "allowed" and proving the correctness of those allowed. Such is a complicated and tedious task, which we shall not undertake in this book. Equation 11.27 contains 13 coefficients, but these were defined in terms of only eight kinetic (or rate) constants. It was the objective of this book to show you how you could think (work) your way through complex enzyme mechanisms enabling you to visualize how changing a mechanism by means of some modifier or by changing the sequence of addition or release of products would change the data plots you might obtain following collection of actual data. I will leave you to the tender mercies of better enzymologists than I should you chose to venture further into the kinetics of enzymology. For myself, I will simply quote Daniel E. Koshland, Jr when he wrote the foreword to the book *Principles of Enzyme Kinetics* by Athel Cornish-Bowden (1976) (coincidentally the year I became a University Assistant Professor): "Those who do not understand and fear mathematics tend either to have excessive admiration for kinetics ('It's too difficult for me.') or excessive contempt ('Kinetics can never prove a mechanism; it can merely disprove one')." I never feared kinetics, but I did not like doing the mathematics—preferring the ethereal means of thinking about the possible mechanisms of an enzyme reaction.

Let's get back to Equation 11.27. Let's see if we can find logic to why the values in the denominator of Equation 11.27 proved to be relevant. It was at this point in my enzymology lecture that I divided the class into small working groups and assigned them specific objectives. However, in this book, it will be a one-on-one effort, and you will have to do all the work yourself (unless your enzymology professor choses otherwise).

Just for the sake of a place to start, let's dissect "c3" in the denominator of Equation 11.27. This coefficient (c3) has no substrate multipliers. It is defined as $c3 = k_{-1}(k_{-2} + k_3)k_4$. If we examine Equation 11.4 we can highlight these steps in the overall enzyme mechanism.

$$E + A \underset{k_{-1}}{\overset{k_1}{\rightleftharpoons}} EA + B \underset{k_{-2}}{\overset{k_2}{\rightleftharpoons}} (EAB/EPQ) \underset{k_{-3}}{\overset{k_3}{\rightleftharpoons}} EQ + P \underset{k_{-4}}{\overset{k_4}{\rightleftharpoons}} Q + E$$

(11.4 repeated)

Coefficient c3, $k_{-1}(k_{-2} + k_3)k_4$ (no substrate multipliers), is illustrated in Equation 11.28:

$$E + A \underset{k_{-1}}{\leftarrow} EA + B \underset{k_{-2}}{\leftarrow} (EAB/EPQ) \overset{k_3}{\rightarrow} EQ + P \overset{k_4}{\rightarrow} Q + E \quad (11.28)$$

Understand what I want you to do? I want you to examine each coefficient from c1 to c13 and look for any "relationships" with respect to their patterns. What enzyme species do they begin and end on? What substrate multipliers are relevant and why? Why are some kinetic constants multipliers while others are only added? Do all the patterns in the denominator coefficients add up to some relevant sum? Can you (we) make any sense out of the patterns (and associated kinetic constants) that were "proved" by mathematical calculations? Remember, understanding comes from the effort, not the final outcome.

I want to emphasize that we will not be solving why Equation 11.27 has those patterns of kinetic constants and substrate multipliers in the denominator. The focus of this book is to show you how changing kinetic constants in enzyme mechanisms due to substrate or modifier binding to an enzyme can shift the way data will appear in data plots. You will be gaining strange data plots from your research, and how you (or whether you can) interpret those data plots will be determined by how well you can reason your way along possible enzyme mechanisms. It should have become obvious to you by now that we have not assigned relative values to any of the kinetic constants in this two-substrate and two-product ordered enzyme mechanism. Depending on how such relative values are assigned (or in actuality occur in real life) the outcome of any velocity value(s) and/or the fraction of each enzyme species present in some steady-state situation of the conversion of substrate to product will change.

While you are doing your part in trying to discern relationships in the coefficients present in the denominator in Equation 11.27, I will continue with my part. The first thing you will notice as I proceed is that I will not be taking the coefficients in the order presented in Equation 11.27. I have already done some "discerning of relationships" in the coefficients present in the denominator of Equation 11.27, and while I will not suggest some "correctness of choice," you may at least see if we are on the same track in our "discerning."

One of the first relationships I noticed were the similarities between coefficient c3 and coefficients c8 and c11. All three combinations of arrows in these three coefficients end on free enzyme (E). They differ, however, in that the combination of arrows in coefficients c8 and c11 also begin on free enzyme (E) and have substrate multipliers whereas the combination of arrows in coefficient c3 begins on EAB and lacks substrate multipliers. Compare Equations 11.29 and 11.30 with Equation 11.28:

$$\text{E} + \text{A} \underset{k_{-1}}{\overset{k_1}{\rightleftarrows}} \text{EA} + \text{B} \underset{k_{-2}}{\overset{k_2}{\rightleftarrows}} (\text{EAB/EPQ}) \underset{k_{-3}}{\overset{k_3}{\rightleftarrows}} \text{EQ} + \text{P} \underset{k_{-4}}{\overset{k_4}{\rightleftarrows}} \text{Q} + \text{E}$$

(11.4 repeated)

Coefficient c8, $k_1k_2(k_3 + k_4)$ (multiplied by ab), is illustrated in Equation 11.29:

$$\text{E} + \text{A} \overset{k_1}{\rightarrow} \text{EA} + \text{B} \overset{k_2}{\rightarrow} (\text{EAB/EPQ}) \overset{k_3}{\rightarrow} \text{EQ} + \text{P} \overset{k_4}{\rightarrow} \text{Q} + \text{E}$$ (11.29)

Coefficient c11, $(k_{-1} + k_{-2})k_{-3}k_{-4}$ (multiplied by pq), is illustrated in Equation 11.30:

$$\text{E} + \text{A} \underset{k_{-1}}{\leftarrow} \text{EA} + \text{B} \underset{k_{-2}}{\leftarrow} (\text{EAB/EPQ}) \underset{k_{-3}}{\leftarrow} \text{EQ} + \text{P} \underset{k_{-4}}{\leftarrow} \text{Q} + \text{E}$$ (11.30)

Have you noticed the similarities between Equation 11.29 and Equation 11.23 as far as the involved steps and kinetic constants (as well as the same substrate multipliers)? The two equations differ only in that in Equation 11.23, $c1 = (k_1 + k_2 + k_3 + k_4)$ and in Equation 11.29, $c8 = k_1k_2(k_3 + k_4)$ with both c1 and c8 being multiplied by substrate concentrations a and b.

If you noticed the similarities between Equations 11.23 and 11.29, you were sure to notice the equivalent similarities between Equations 11.24 and 11.30 as far as the involved steps and kinetic constants (as well as the same substrate multipliers). These latter two equations differ only in that in Equation 11.24, $c2 = (k_{-1} + k_{-2} + k_{-3} + k_{-4})$ and in Equation 11.30, $c11 = (k_{-1} + k_{-2})k_{-3}k_{-4}$ with both c2 and c11 being multiplied by substrate concentrations p and q.

There are only two additional patterns of four arrows in the coefficients in the denominator of Equation 11.27 and these are associated with coefficients c4 and c7 illustrated below as Equations 11.31 and 11.32.

$$\text{E} + \text{A} \underset{k_{-1}}{\overset{k_1}{\rightleftarrows}} \text{EA} + \text{B} \underset{k_{-2}}{\overset{k_2}{\rightleftarrows}} (\text{EAB/EPQ}) \underset{k_{-3}}{\overset{k_3}{\rightleftarrows}} \text{EQ} + \text{P} \underset{k_{-4}}{\overset{k_4}{\rightleftarrows}} \text{Q} + \text{E}$$

(11.4 repeated)

Coefficient c4, $k_1(k_{-2} + k_3)k_4$ (multiplied by a), is illustrated in Equation 11.31:

$$E + A \xrightarrow{k_1} EA + B \underset{k_{-2}}{\longleftarrow} (EAB/EPQ) \xrightarrow{k_3} EQ + P \xrightarrow{k_4} Q + E \quad (11.31)$$

Coefficient c7, $k_{-1}(k_{-2} + k_3)k_{-4}$ (multiplied by q), is illustrated in Equation 11.32:

$$E + A \underset{k_{-1}}{\longleftarrow} EA + B \underset{k_{-2}}{\longleftarrow} (EAB/EPQ) \xrightarrow{k_3} EQ + P \underset{k_{-4}}{\longleftarrow} Q + E \quad (11.32)$$

Both the mechanisms shown in Equations 11.31 and 11.32 begin (arrows originate) on both E and EAB/EPQ and end (arrows terminate) on E. However, Equation 11.31 also ends on EA, and Equation 11.32 also ends on EQ. Thus, in some respects the two mechanisms are similar with the exception of one essentially representing a "forward" reaction (with a backward reaction ending on EA) and binding substrate A to free enzyme E, and the latter representing a backward reaction (with a forward reaction ending on EQ) and binding substrate Q to free enzyme E.

Further, Equation 11.31 differs from Equation 11.29 in substituting k_{-2} in place of k_2, and Equation 11.32 differs from Equation 11.30 in substituting k_3 in place of k_{-3}. However, the combinations of kinetic constants between the mechanisms in Equations 11.29 and 11.31 are dramatically different, $k_1 k_{-2}(k_3 + k_4)$ versus $k_1(k_{-2} + k_3)k_4$, as are the combinations of kinetic constants between the mechanisms in Equations 11.30 and 11.32, $(k_{-1} + k_{-2})k_{-3}k_{-4}$ versus $k_{-1}(k_{-2} + k_3)k_{-4}$, signaling a disconnect between the visible differences in mechanisms versus the kinetic constants, ultimately proven by complicated (mostly computer-driven) mathematical calculations designed to allow and not allow specific patterns of arrows and kinetic constants.

These were the only coefficients to contain four kinetic constants (and associated arrows) in mechanism patterns. The remaining six coefficients (c5, c6, c9, c10, c12, and c13) contain only three kinetic constants (and associated arrows) in the mechanism patterns. Again, just looking for pattern similarities we observe that coefficients c5 and c6 are "similar," coefficients c9 and c10 are similar, and coefficients c12 and c13 are similar.

Let's examine these "three arrow" (three kinetic constants) mechanisms that constitute the final six coefficients in the denominator of Equation 11.27. Let's begin with coefficients c5 and c6 illustrated with arrows and kinetic constants in Equations 11.33 and 11.34. However, since I want you to be able to quickly compare the additional coefficients c9, c10, c12, and c13 I propose to just include them in the text one after the other without any written comments.

$$E + A \underset{k_{-1}}{\overset{k_1}{\rightleftharpoons}} EA + B \underset{k_{-2}}{\overset{k_2}{\rightleftharpoons}} (EAB/EPQ) \underset{k_{-3}}{\overset{k_3}{\rightleftharpoons}} EQ + P \underset{k_{-4}}{\overset{k_4}{\rightleftharpoons}} Q + E$$

(11.4 repeated)

Coefficient c5, $k_2k_3k_4$ (multiplied by substrate b), is illustrated in Equation 11.33:

$$EA + B \overset{k_2}{\rightarrow} (EAB/EPQ) \overset{k_3}{\rightarrow} EQ + P \overset{k_4}{\rightarrow} Q + E \qquad (11.33)$$

Coefficient c6, $k_{-1}k_{-2}k_{-3}$ (multiplied by substrate p), is illustrated in Equation 11.34:

$$E + A \underset{k_{-1}}{\leftarrow} EA + B \underset{k_{-2}}{\leftarrow} (EAB/EPQ) \underset{k_{-3}}{\leftarrow} EQ + P \qquad (11.34)$$

Coefficient c9, $k_1k_{-2}k_{-3}$ (multiplied by substrates ap), is illustrated in Equation 11.35:

$$E + A \overset{k_1}{\rightarrow} EA + B \underset{k_{-2}}{\leftarrow} (EAB/EPQ) \underset{k_{-3}}{\leftarrow} EQ + P \qquad (11.35)$$

Coefficient c10, $k_2k_3k_{-4}$ (multiplied by substrates bq), is illustrated in Equation 11.36:

$$EA + B \overset{k_2}{\rightarrow} (EAB/EPQ) \overset{k_3}{\rightarrow} EQ + P \underset{k_{-4}}{\leftarrow} Q + E \qquad (11.36)$$

Coefficient c12, $k_1k_2k_{-3}$ (multiplied by substrates abp), is illustrated in Equation 11.37:

$$E + A \overset{k_1}{\rightarrow} EA + B \overset{k_2}{\rightarrow} (EAB/EPQ) \underset{k_{-3}}{\leftarrow} EQ + P \qquad (11.37)$$

Coefficient c13, $k_2k_{-3}k_{-4}$ (multiplied by substrates bpq), is illustrated in Equation 11.38:

$$EA + B \overset{k_2}{\rightarrow} (EAB/EPQ) \underset{k_{-3}}{\leftarrow} EQ + P \underset{k_{-4}}{\leftarrow} Q + E \qquad (11.38)$$

I don't wish to labor the point that I undertook this exercise with you not to demonstrate why these particular combinations of kinetic constants and substrate multipliers turned out to be "proven" by other enzymologists, but I do emphasize that the exercise was important in getting you to work your way through this process. What did you come away with from performing this exercise? Did you consider what each mechanism began and ended on with respect to enzyme species? Did you consider which substrates were consumed in specific mechanisms? Did you consider which products were released in specific mechanisms? Was your curiosity piqued sufficient for you to want to solve why the specific combinations of kinetic constants and substrate multipliers were "proven" by mathematical calculations?

Let's see if we can gain further information from this ordered two-substrate and two-product enzyme mechanism. You have probably noticed by now that we have not assigned relative values to any of the kinetic constants in Equation 11.27. We have stipulated that rate equation derivation for Equation 11.27 occurs under conditions of "saturation" by all substrates, under equilibrium conditions (meaning we are not trying to derive some rate equation under initial velocity conditions where product is not present), and that the mechanism is under some steady state of flux. A "steady state of flux" is not the same as steady-state assumption and thus we are free to stipulate as we did with one-substrate and one-product mechanisms in earlier chapters that the enzyme mechanism illustrated in Equation 11.4 can be defined as occurring under quasi-equilibrium assumptions.

Assignment of Values to Kinetic Constants: Ordered Two-Substrate/Two-Product Mechanism

Up to now, we have considered the enzyme mechanism illustrated in Equation 11.4 for purposes of understanding the basis for calculating and deriving a rate equation. In the derivation of some rate equation, the relative values of the various kinetic constants do not factor into the derivation. However, for our purposes, the relative values of the various kinetic constants are extremely important in that they determine how substrates are converted to products as well as how these conversions might be modified as the needs of cellular metabolism change. Let's look again at Equation 11.4:

$$E + A \underset{k_{-1}}{\overset{k_1}{\rightleftharpoons}} EA + B \underset{k_{-2}}{\overset{k_2}{\rightleftharpoons}} (EAB/EPQ) \underset{k_{-3}}{\overset{k_3}{\rightleftharpoons}} EQ + P \underset{k_{-4}}{\overset{k_4}{\rightleftharpoons}} Q + E$$

(11.4 repeated)

Let's assign some relative values to each of the kinetic constants in Equation 11.4. In addition, let's stipulate that we will make such assignments with the objective of setting up these relative values to achieve quasi-equilibrium assumptions for the mechanism. Achieving quasi-equilibrium assumptions for Equation 11.4 will permit us to consider modification events that could be associated with an ordered two-substrate and two-product enzyme mechanism.

Let's stipulate the following relative values for the kinetic constants in Equation 11.4:

$$k_1 > k_{-1}; k_2 > k_{-2}; k_1 = k_2; k_1 \gg k_3; k_3 = k_4; k_3 > k_{-3}; k_4 > k_{-4}; k_{-4} = k_{-3}.$$

These relative values for the kinetic constants in Equation 11.4 would equate to the following quasi-equilibrium conditions for the enzyme mechanism:

1) That substrate A will bind to free enzyme E equal to the binding of substrate B to the EA enzyme species forming the EAB/EPQ enzyme species.
2) The EAB/EPQ enzyme species will break down to the EQ enzyme species releasing product P faster than the EQ enzyme species will break down to the EA enzyme species releasing substrate B.
3) The EQ enzyme species will break down to free enzyme E releasing product Q at the same rate that the EAB/EPQ enzyme species will break down to the EQ enzyme species releasing product P.
4) The forward reaction will occur faster than the reverse reaction.

There are, of course, additional points that can be made with respect to the additional relationships (specified and not specified) that occur when relative values are assigned to all eight kinetic constants in Equation 11.4. However, the four points made above are the most relevant for our further discussion.

Although not explicitly made, the K_m values for substrates A and B are essentially equivalent (if not equal), as are the K_m values for substrates P and Q in the reverse mechanism. However, it is also implied that the K_m values for the substrates P and Q are greater (requires higher concentrations) than the K_m values for the substrates A and B, and the rate of conversion of substrates A and B (at saturating concentrations) to products P and Q will be greater than the rate of conversion of substrates P and Q (again at saturating concentrations) to products A and B.

If we determine the rate of conversion of substrate B (let's go simple first) to product(s) P and Q via Equation 11.4 at increasing concentrations of substrate B (where the concentration of substrate A is present in saturating concentrations), we will obtain a velocity versus concentration data plot very similar to that shown as Figure 3.1 reproduced here as Figure 11.1.

We will obtain a similar velocity versus substrate concentration data plot were we to measure the rate of conversion of substrate "A" (as illustrated in Equation 11-3) to products "P" and "Q" in the presence of saturating concentrations of substrate "B". In addition, although I have written this book with the focus on the forward reactions of given enzyme mechanisms, we could easily perform similar enzyme assays where we measure rates of conversion of substrates to products via the reverse route of the enzyme mechanism illustrated in Equation 11-3.

Given the above comments, you should begin to realize that we can undertake evaluations of the roles of various kinds of enzyme modifiers where such modifiers interact with some aspect of the enzyme mechanism illustrated by Equation 11.4. However, in order to do this evaluation in a manner that is the

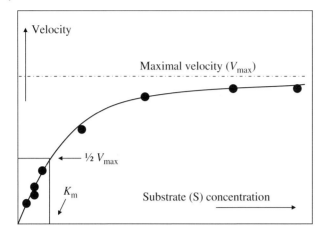

Figure 11.1 An expected velocity versus substrate concentration data plot for substrate B illustrated in Equation 11.4 in the presence of saturating concentrations of substrate A.

least complicated, I propose to modify Equation 11.4 to the mechanism shown in Equation 11.39:

$$E + A \underset{k_{-1}}{\overset{k_1}{\rightleftarrows}} EA + B \underset{k_{-2}}{\overset{k_2}{\rightleftarrows}} (EAB/EPQ) \xrightarrow{k_3} EQ + P \xrightarrow{k_4} Q + E \quad (11.39)$$

All I have done with Equation 11.4 to turn it into Equation 11.39 is make steps 3 and 4 of the enzyme mechanism irreversible. This change simply means we will not need to consider the reverse reaction in the conversion of substrates (A and B) to products (P and Q) in subsequent assessments of the roles of modifiers (inhibitors and/or activators) in the two-substrate to two-product enzyme mechanisms.

I would like to make one additional change to Equation 11.4 (or if you wish Equation 11.39) to create Equation 11.40:

$$E + A \underset{k_{-1}}{\overset{k_1}{\rightleftarrows}} EA + B \underset{k_{-2}}{\overset{k_2}{\rightleftarrows}} (EAB/EPQ) \xrightarrow{k_3} EQ + P \quad (11.40)$$

You will note in Equation 11.40 that this enzyme mechanism is essentially an irreversible two-substrate and one-product enzyme mechanism, and if we specify that the sequence of addition of substrates is ordered, we have a pretty simple enzyme mechanism where we could (at this time) assign relative values to the kinetic constants to keep ourselves under quasi-equilibrium assumptions. However, you will have noticed that we have a "problem" in Equation 11.40. We end the equation with the enzyme species EQ rather than E, and given this event, we will soon exhaust the amount of free enzyme (E). The easiest way to deal with this is once again to modify the equation (Equation 11.40) to yield a new equation, 11.41, where we (I) simply eliminate the second substrate Q yielding a pretty simple enzyme mechanism where two substrates are irreversibly converted to a single product (which is, of course somewhat analogous to the D-arabinoheptulosonate-7-phosphate synthetase enzyme we have seen before) and looks a lot like Equation 10.1.

$$E + A \underset{k_{-1}}{\overset{k_1}{\rightleftharpoons}} EA + B \underset{k_{-2}}{\overset{k_2}{\rightleftharpoons}} (EAB/EPQ) \overset{k_3}{\longrightarrow} E + P \tag{11.41}$$

So, why you might wonder, did I take you on what might appear to be a circuitous route through a two-substrate/two-product mechanism and complicated rate equation derivation in Chapter 11, only to bring you back to a two-substrate and one-product mechanism that we dealt with in Chapter 10? If you look at the means by which modifiers modify some enzyme conversions of substrates to product in Chapter 10, you will note that I created mechanisms where the modifiers only modified free enzyme ("E"). Refer to Equation 10.9 as an example. Given that modifiers can bind to enzyme–substrate complexes as readily (easily) as they can bind to free enzyme, do we need now to consider a new round of enzyme mechanisms in the way we considered modification of enzyme mechanisms in Chapter 10, but in a more complicated manner?

12

Modification of Enzyme Mechanisms

The Next Generation

There are many ways to study enzymes and their mechanisms of action, but certainly one of the best, and the one that must be the final arbiter, is the study of the kinetics of the catalytic reaction itself. An enormous amount of information can be obtained from such studies, but only if one understands the subtle nuances of the theory involved can one properly set up and carry out the necessary experiments.

W.W. Cleland in the Foreword to *Enzyme Kinetics* by K.M. Plowman
(McGraw-Hill Publishers, 1972)

I saved the quotation shown at the top of this chapter for several reasons. Firstly, I received my PhD in 1972, and it was the work of Cleland that I most admired from my early studies of enzymes in college. Secondly, I think the above quote reflects an important philosophy that led to this book. If you can't understand and elucidate the theories regarding how enzymes can be modified and/or interact with substrates, you can learn to derive all the kinetic equations in the world, yet fail to understand.

You will recall that at the end of Chapter 11 I left you with an enzyme mechanism where two substrates were being converted to one product. Equation 11.41 is repeated here as Equation 12.1:

$$E + A \underset{k_{-1}}{\overset{k_1}{\rightleftharpoons}} EA + B \underset{k_{-2}}{\overset{k_2}{\rightleftharpoons}} (EAB/EP) \xrightarrow{k_3} E + P \qquad (12.1)$$

If we stipulate that $k_1 \geq k_{-1}$, $k_2 \geq k_{-2}$, $k_1 = k_2$, and $k_3 << k_1$ (and conversely k_2), we will have satisfied the objective of rendering Equation 12.1 (or Equation 11.41 repeated) as describing an enzyme reaction as a two-substrate/one-product, irreversible enzyme reaction that operates under quasi-equilibrium assumptions.

Unlike in Chapter 10 where we had only the option(s) of modifiers binding to free enzyme, it must be almost irresistible for you to begin to think about how Equation 12.1 might "look" if one or more modifiers were able to bind to the EA

Enzyme Regulation in Metabolic Pathways, First Edition. Lloyd Wolfinbarger.
© 2017 John Wiley & Sons, Inc. Published 2017 by John Wiley & Sons, Inc.

enzyme complex rather than to free enzyme (E), such as was illustrated in Equation 10.7, which is repeated here as Equation 12.2:

$$\text{Ma1E} + A \underset{k_{-7}}{\overset{k_7}{\rightleftharpoons}} \text{Ma1EA} \underset{B \downarrow k_{-8}}{\overset{B \rceil k_8}{\rightleftharpoons}} \text{Ma1EAB} \xrightarrow{k_9} \text{Ma1} + E + P$$

$$\text{Ma2E} + A \underset{k_{-10}}{\overset{k_{10}}{\rightleftharpoons}} \text{Ma2EA} \underset{B \downarrow k_{-11}}{\overset{B \rceil k_{11}}{\rightleftharpoons}} \text{Ma2EAB} \xrightarrow{k_{12}} \text{Ma2} + E + P$$

$$E + A \underset{k_{-1}}{\overset{k_1}{\rightleftharpoons}} EA \underset{B \downarrow k_{-2}}{\overset{B \rceil k_2}{\rightleftharpoons}} EAB \xrightarrow{k_3} E + P$$

$$\text{MiE} + A \underset{k_{-4}}{\overset{k_4}{\rightleftharpoons}} \text{MiEA} \underset{B \downarrow k_{-5}}{\overset{B \rceil k_5}{\rightleftharpoons}} \text{MiEAB} \xrightarrow{k_6} \text{Mi} + E + P \qquad (12.2)$$

(Diagram also includes vertical couplings: Ma1 with k_{13}, k_{-13}; Ma2 with k_{14}, k_{-14}; Mi with k_{15}, k_{-15}.)

Let's reconfigure Equation 12.2/10.7 (creating Equation 12.3), but this time let's illustrate the binding of modifiers (both inhibitors and activators of course) to the EA enzyme complex rather than to free enzyme (E) as was shown in Equation 12.2.

However, before we get into the enzyme mechanisms to be illustrated in Equation 12.3, let me re-emphasize that the intent here is that all enzyme mechanisms that convert substrates to product (whether modified or not modified) operate under quasi-equilibrium assumptions. If you can accept this stipulation, we (I) won't have to define the relationships between the kinetic constants in the enzyme mechanisms that are modified. However, it will be helpful to you if you jot the diagrams from Equation 12.3 to a separate sheet of paper and define the relationships of all rate constants to each other. This task is not requested lightly (or trivially) in that as we begin to consider all possible combinations of modification(s) that can occur with even this simplistic set of enzyme mechanisms, you will find yourself needing to manipulate relative values (i.e., relationships) of differing rate constants in order to maintain a given modifier as either an activator or an inhibitor of the nonmodified enzyme mechanism. In short, the opportunity for confusion rises exponentially and things will get complicated beyond your willingness to stay with me.

Let's examine Equation 12.3 by simply comparing it to Equation 12.2 (10.7 repeated).

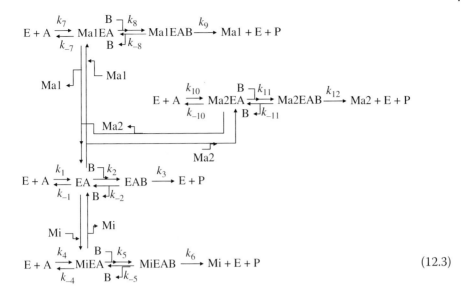

$$(12.3)$$

I think it is obvious that the only real mechanistic differences between Equations 12.3 and 12.2 lie in the enzyme form that the modifiers bind to, free enzyme (E) versus substrate (A)-bound enzyme (EA).

However, the real complicating aspects of Equations 12.3 and 12.2 lie in how and which kinetic constants are changed in the modification being driven by the respective modifiers (and there are three of them).

Remember that each enzyme mechanism shown in both Equations 12.2 and 12.3 operate under quasi-equilibrium assumptions, and thus the rate-limiting conversion of the two substrates to the one product is always that step where product is released (the K_{cat} term frequently used in other text-books on enzymology), for example, k_3 in the nonmodified mechanisms in both equations. It is also important to re-emphasize that in the presence of modifiers, all four mechanisms will presumably be metabolizing the two substrates into product, and thus some observed overall apparent K_m and apparent V_{max} values you might obtain from experimentation will represent the sum of values for the apparent K_m and apparent V_{max} values for all four such enzyme mechanisms.

However, and there is always a "however" in enzymology, those mechanisms illustrated in Equations 12.2 and 12.3 represent only two of some larger possible combinations of equations that can be generated. I will list a few such possible combinations below, but please remember that the list I provide is not complete.

1) Equations 12.2 and 12.3 show modifier Ma2 binding to free enzyme (E) and the enzyme substrate (E) complex, respectively. Suppose the binding of modifier Ma1 differed from that shown in these two equations and Ma1 bound to free enzyme (E) in Equation 12.3 (rather than to EA) and to the enzyme–substrate complex (EA) in Equation 12.2 (rather than to E)?

2) Consider that each of the four possible enzyme mechanisms described in item 1 carries the additional option(s) of flipping how the modifier (Mi) binds to either free enzyme (E) or substrate-bound enzyme (EA) in Equations 12.2 and 12.3.

Those of you who are more enthusiastic will figure out all such possible combinations, but rather than me dwelling on these possible combinations I prefer to move on to the next possible combinations of enzyme mechanisms that we can discuss based on which rate constants appear to be changed by a modifier.

As a *temporary measure*, I would like to stipulate that in Equations 12.2 and 12.3, the rate constants k_3, k_6, k_9, and k_{12} are identical (i.e., $k_3 = k_6 = k_9 = k_{12}$). This temporary measure, of course, means that the presence of modifiers will not change the apparent V_{max} in the conversion of the two substrates to the one product, and thus I am limiting the role(s) of modifiers to changing the apparent K_m values of either free enzyme (E) or substrate-bound enzyme (EA) for either (or both) substrate A or substrate B. We start simple and move to more complex.

Once again, I propose to provide a short (albeit incomplete) list and some possible changes to rate constants that the enzyme modifiers can affect (and, of course, leave the complete list to those of you enthusiastic enough to figure them all out). Consider the following available based on the mechanisms illustrated in Equations 12.2 and 12.3:

1) Consider that the binding of the modifiers Ma1, Ma2, and Mi to either the free enzyme (E) or the substrate-bound enzyme (EA) modify only the apparent K_m values for the binding of substrate A to free enzyme E and the binding of substrate B to substrate-bound enzyme EA.
2) Consider that the binding of the modifier Ma1 to the free enzyme (E) modifies both the apparent K_m values for the binding of substrate A to free enzyme E *and* the binding of substrate B to substrate-bound enzyme EA.
3) Consider that the binding of the modifier Ma2 to the free enzyme (E) modifies both the apparent K_m values for the binding of substrate A to free enzyme E *and* the binding of substrate B to substrate-bound enzyme EA.
4) Consider that the binding of the modifier Mi to the free enzyme (E) modifies the apparent K_m value for the binding of substrate B to substrate-bound enzyme EA.

And, you have, of course, identified some additional considerations in the above list not readily describable because the possible enzyme mechanisms are not illustrated in Equations 12.2 and 12.3. Consider a few of the possible additional options:

1) Consider that all three modifiers could bind to free enzyme (E), prior to the binding of substrate A to free enzyme E, and modify the apparent K_m value of enzyme E for substrate A based on the possible combinations of modifier-bound enzyme. For example, Ma1/Mi/E, Ma2/Mi/E, Ma1/Ma2/E, and Ma1/Ma2/Mi/E.

2) Consider that all three modifiers could bind to substrate-bound enzyme (EA), prior to the binding of substrate B to the substrate-bound enzyme EA, and modify the apparent K_m value of EA for substrate B. For example, Ma1/Mi/EA, Ma2/Mi/EA, Ma1/Ma2/EA, and Ma1/Ma2/Mi/EA.

I have belabored this point, and challenged you to work out additional possible mechanisms not illustrated in Equations 12.2 and 12.3 because I think it important that as you consider possible mechanisms (or combinations of mechanisms) that might fit your data, that you take time to work yourself through all options you can think of that will even come close to fitting the data you obtain from experimentation. Remember the admonition (if I can call it an "admonition") of Cleland I quoted at the beginning of this chapter: "...but only if one understands the subtle nuances of the theory involved...."

Finally (at least for this part of this chapter), you will have noted in the narrative above that I only made reference to changing some apparent K_m value by some modifier (or combination of modifiers) and did not attempt to clarify whether the modification of some apparent K_m could be attributable to apparent changes to either (or both) of the rate constants constituting the apparent K_m under discussion (remember that under our terms of discussion, the enzyme mechanisms operate under quasi-equilibrium assumptions and therefore the rate constants k_3, k_6, k_9, and k_{12} do not contribute (or contribute "minimally") to the K_m value of an enzyme for substrate. This means that any modification event changes only (or results in new) association and/or dissociation rate constants for any given enzyme mechanism (modified or not modified by a modifier). This event is of special importance in considering the association and dissociation rate constants illustrated as occurring between the enzyme forms EA and EAB. The association rate constant will, of course, be multiplied by some factor associated with the concentration of substrate B and the concentration of the enzyme–substrate complex EA. The dissociation rate constant will, of course, be multiplied by some factor associated with the concentration of the enzyme–substrate complex EAB, but not some substrate concentration.

Role of the Binding Constants of Modifiers

So far I have focused on the value(s) of possible enzyme mechanisms (modified and not modified) as well as how modifiers may result in differing kinetic constants for the similar steps in the unmodified and modified mechanisms for the conversion of two substrates to one product. It is now time to consider the role of the association and dissociation rate constants (and concentration(s) of modifiers) associated with the binding of a modifier (or multiple modifiers) to some form of the enzyme involved in the enzymatic conversion. Although I fully appreciate that you are tired of seeing Equation 10.7/12.2, I am reproducing it here again in order that you may review those kinetic constants involved in the binding of modifiers to free enzyme (E). However, you should not forget that these same modifiers could just as easily bind to the enzyme–substrate complex EA as illustrated in Equation 12.3.

$$MalE + A \underset{k_{-7}}{\overset{k_7}{\rightleftarrows}} MalEA \underset{B\,\downarrow k_{-8}}{\overset{B\,\rceil k_8}{\rightleftarrows}} MalEAB \xrightarrow{k_9} Mal + E + P$$

$$Ma2E + A \underset{k_{-10}}{\overset{k_{10}}{\rightleftarrows}} Ma2EA \underset{B\,\downarrow k_{-11}}{\overset{B\,\rceil k_{11}}{\rightleftarrows}} Ma2EAB \xrightarrow{k_{12}} Ma2 + E + P$$

$$E + A \underset{k_{-1}}{\overset{k_1}{\rightleftarrows}} EA \underset{B\,\downarrow k_{-2}}{\overset{B\,\rceil k_2}{\rightleftarrows}} EAB \xrightarrow{k_3} E + P$$

$$MiE + A \underset{k_{-4}}{\overset{k_4}{\rightleftarrows}} MiEA \underset{B\,\downarrow k_{-5}}{\overset{B\,\rceil k_5}{\rightleftarrows}} MiEAB \xrightarrow{k_6} Mi + E + P$$

(connecting modifiers: Ma1, Ma1 with k_{13}, k_{-13}; Ma2, Ma2 with k_{14}, k_{-14}; Mi, Mi with k_{15}, k_{-15})

(12.2 repeated)

As you can see from Equation 12.2 (repeated), we have been considering a complex enzyme mechanism involving three modifiers (Ma1, Ma2, and Mi) binding to free enzyme E. A simplistic way to think about this option would be to assume that free enzyme E has four binding sites. One binding site for the binding of substrate A and three sites for the binding of the three modifiers. This simplistic option more or less stipulates that binding of one modifier to free enzyme (E) precludes the binding of the other two modifiers, yielding four mechanisms that can be involved in the conversion of two substrates (A and B) to one product (P). This option also stipulates that in the presence of all three modifiers, the numbers of free enzyme molecules that will be distributed among these four mechanisms will be controlled by the relative values of the association rate constants k_{13}, k_{14}, and k_{15} and the dissociation rate constants k_{-13}, k_{-14} and k_{-15}. There are, of course, the additional factors of relative concentrations of each of these modifiers to be dealt with, but initially we will seek to deal with an option where the concentrations of all three modifiers are equivalent and at sufficient concentrations to saturate free enzyme (E) with any given modifier (and that the amount of each modifier present is unlimited).

There are way too many options by which the overall conversion of substrate(s) to product can be altered, and we will only discuss a small number of such options with the expectation that you will readily be able to come up with additional such options as your needs dictate.

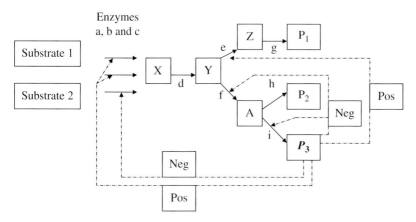

Figure 12.1 Diagram of the aromatic amino acid biosynthetic pathway showing the enzymatic steps in the pathway at which tyrosine (P_3) modifies enzymatic activity via positive and negative regulatory events [Figure 4.5 repeated].

A good quick first option to consider would be if we stipulated that $k_{13} = k_{14} = k_{15}$ and $k_{-13} = k_{-14} = k_{-15}$. This option would mean that all three modifiers have the same K_{eq} value ("affinity constant" for the binding of modifier) and in the presence of saturating concentrations of modifiers, there would be no conversion of substrates to product by free enzyme E. Only the modified pathways would be involved in the conversion of substrates to product, and the apparent K_m value and apparent V_{max} you would calculate through experimentation would be the average of the three K_m values and the average of the three V_{max} values of the modified mechanisms. Given that these "average" values for K_m and/or V_{max} will be different from the apparent K_m and V_{max} values for the nonmodified enzyme mechanism, you can expect a change in the rates of conversion of the two substrates to the product. Whether the change is an apparent inhibition or activation will depend on how you establish the relationships of all the "analogous" rate constants. Understand how I want you to think this through?

Remember, Mi is supposed (at least in our prior discussions) to be an inhibitor of the enzymatic conversion of substrates A and B to product P. Thus, we have an option where we can stipulate that k_{15} is greater than either k_{13} and/or k_{14} and begin to see an option where more free enzyme E is shifted to the MiE form of enzyme than to either Ma1E or Ma2E. However, we have not stipulated any relationships between the three dissociation rate constants (k_{-13}, k_{-14}, and k_{-15}). Can we let them all be equal to each other or do we need to make k_{-15} smaller than both k_{-13} and k_{-14} to exaggerate the shifting of free enzyme toward the inhibited enzyme mechanism? Have you begun to realize the tenuous nature of some effort you might chose to undertake in working on additional options?

Let's go back to our aromatic amino acid biosynthetic pathway and specifically the DAHP synthetase enzyme and how the various "end products" of that pathway serve to regulate aromatic amino acid biosynthesis. Let's review Figures 4.5, 4.6, and 4.7, reproduced here as Figures 12.1, 12.2, and 12.3.

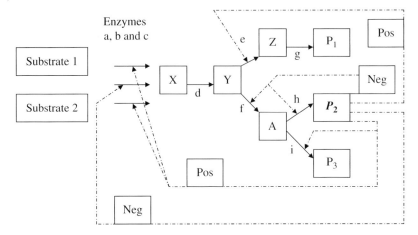

Figure 12.2 Diagram of the aromatic amino acid biosynthetic pathway showing the enzymatic steps in the pathway at which phenylalanine (P$_2$) modifies enzymatic activity via positive and negative regulatory events [Figure 4.6 repeated].

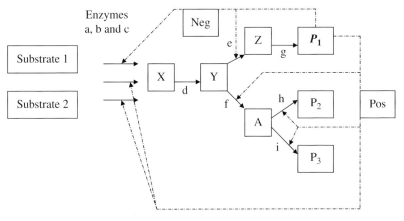

Figure 12.3 Diagram of the aromatic amino acid biosynthetic pathway showing the enzymatic steps in the pathway at which tryptophan (P$_1$) modifies enzymatic activity via positive and negative regulatory events [Figure 4.7 repeated].

You have probably already realized the point I wish to now make. In the conversion of erythrose-4-phosphate and phosphoenolpyruvate to D-arabinoheptulosonate-7-phosphate by the enzymes D-arabinoheptulosonate-7-phosphate synthetase, there is an "s" at the end of the word "enzyme" in this sentence. Therefore, at this enzymatic step you will have to deal with combinations of three sets of Equations 12.2, 10.9, and any and all other possible combinations of enzyme mechanisms should modifiers exhibit some mixture of binding to free enzyme E or to substrate-bound enzyme EA. My initial question when I first began to think about this was: "Why would a living cell evolve such a convoluted enzyme step?" (If you look at enzyme e in each of these three figures, you will notice that this enzymatic step is similar to each of the three isozymes of the DAHP synthetase, but to date there appears to be only one isozyme of enzyme e in this pathway.)

There are, of course, several way to think about some answer to the "why" above. Enzymatic activity is regulated in multiple ways. The cell can regulate how many molecules of a particular enzyme are produced through regulation of both transcription of genes and translation of messenger RNAs. If a cell were to dramatically reduce transcription and/or translation as a means of reducing levels of the enzyme DAHP synthetase (and there were not three isozymes, i.e., genes, for this enzyme) the ability of a cell to synthesize aromatic amino acids would be dramatically reduced (as well as several other metabolites derived from intermediates in this pathway). Thus, this approach to regulation is pretty much an entirely different approach to the regulation of enzymatic activity than the one we have been discussing. There is, of course, also degradation of enzyme(s) during the course of normal cellular turnover, but this means of regulation is also very different to the one we have been discussing. One reason for not covering these modes of enzyme regulation in this book has to do with a lack of effect(s) on the kinds of data plots you can expect to gain from experimentation, and this book is all about learning to think about how to conduct experiments that will tell you about how an enzyme manages the conversion of some substrate to some product.

By now you have probably begun to wonder whether my nature is just to be perverse or whether I am just "wordy" and like to hear myself lecture to you in written format. I like to think neither. I like to think that you are, by now, beginning to understand the nature of what it takes to think your way through options when it comes to trying to develop some mechanism or combinations of mechanisms to explain the experimental data you will obtain from your experimentation. Those of you who become hard core enzymologists will have computer programs to plug your values into and see, via some visualized data plot, your graphic data. The rest of you, those of you who pull wings off of flies, will need to develop a sixth sense that will aid you in visualizing how changing some relationship(s) of a given rate constant to other rate constants, or adding some option for some modification event to add mechanisms to the reaction, might make sense. It is for this latter reason that I have been verbose, repetitive, and my lecture is in a written format. In a classroom lecture, I was able to watch the responses and facial expressions of the students. In this book, I have neither opportunity and can only hope that you will forgive me for the tedious nature of my writing.

I saved the penultimate chapter in Part II in anticipation that you would already know much of what I talked about in this chapter and that you might become bored if I put the contents of this chapter early on. I wanted you to get to the part where you think about and hopefully argue about what I have provided. If this book has a fault, it is that it has limited you to speculations of interactions based on quasi-equilibrium assumptions, and to my knowledge, no enzyme approximates quasi-equilibrium assumptions. Most enzymes operate under some form of steady-state assumptions (briefly, this is where k_2 is not much much less than k_1 and/or k_{-1}) and where the most probable mechanism is infinitely more complex than the simple mechanisms presented to you in this book. My objective here has been to get you to think, analyze, challenge, question, object, and argue with yourself and me. I appreciate that it will be difficult to do these things with me, but I hope that you will always do these things with yourself and those who try to teach you. Such is the basis of acquiring knowledge.

13

What Are These "Rate Constants" We Have Been Dealing With?

All through this text we have been changing the ways in which an enzyme facilitates some catalytic event in changing one molecule to another molecule. We have mostly dealt with the use of "modifiers" binding at some site on the enzyme that changes some kinetic (or rate) constant such that either the V_{max} or K_m of that enzyme changes. The question, however, is: Does the modifier actually change a rate constant in the modified state of the enzyme or does it simply replace a specific rate constant for another rate constant in the modified state of the enzyme? Perhaps an even more important question is: What is a rate (kinetic) constant?

Kinetic (rate) constants are generally described as approximating some number of moles per second—meaning, for example, that some number of moles of a substrate molecule can bind to some binding site of an enzyme (per second) resulting in what I have been describing as the "substrate–enzyme complex," or some substrate can be changed to "product" and released by the enzyme as some number of moles of product molecule (again, per second). By now, of course, you will realize that very little in enzymology is real or finite, and with all such things a "rate constant" is probably not constant. A "rate constant" you calculate through experimentation will most likely represent some "average" value, with a range of such rate constant values varying over some narrow or wide range. This variation depends on a number of confounding things that can (and will) influence the rate at which a substrate molecule can (will) bind to free enzyme to form an enzyme–substrate complex (in our present example) that will be converted to product that can be released along with free enzyme. It will be good at this point to remind you that the purpose of this text is not to teach you how to calculate a rate constant, but rather to cause you to think about calculating rate constants when everything around you is trying to make you crazy.

So, having given you a simple definition of a rate (kinetic) constant (moles per second), just how do you think about calculating a rate constant for some specific event in an enzyme-catalyzed conversion of one molecule to another molecule? Remember that this conversion of one molecule to another molecule will occur without the presence of the enzyme so we will want to be able to separate simple chemical conversion events from enzyme-mediated conversion events.

Just now, it would be great to have some visuals showing a substrate molecule vibrating while it rolls and tumbles amongst solvent (let's keep it simple and say

Enzyme Regulation in Metabolic Pathways, First Edition. Lloyd Wolfinbarger.
© 2017 John Wiley & Sons, Inc. Published 2017 by John Wiley & Sons, Inc.

water) molecules moving erratically toward (we hope toward) an equally vibrating, rolling/tumbling enzyme molecule, which (hopefully) is also moving (again erratically) toward that lone substrate molecule. Now, try to visualize the ordered/disordered nature of the water molecules such that the substrate and enzyme have each contributed to some degree of "order" of the water molecules in close proximity to them, but those water molecules lying beyond the sphere of influence of the substrate molecule/enzyme molecule have a less ordered arrangement. If you are still with me, you are probably now thinking about what is probably an enormous difference in sizes between the substrate molecule and the enzyme molecule, and are beginning to wonder how the substrate molecule gets close to the enzyme's substrate-binding site. You might want to ignore, just for now, whether the substrate molecule (even if it arrives in close proximity to the substrate-binding site on the enzyme molecule) will be trying to "enter" some substrate-binding site in some proper orientation such that the bond or group on the substrate to be changed comes into proximity to those bonds or groups on the enzyme that will facilitate the anticipated change. Does this enzyme-mediated conversion of substrate to product work simply as a statistical probability event (meaning most of the time this coming together in some proper circumstance is an "accident") or might there be other factors working in this process about which we are ignorant? What other factors can you think of at this point?

Try to stay in the right side of your brain for now. The following is not about the analytical aspect of this effort, but rather is about you trying to think about factors that might (may? can?) influence some enzyme-facilitated conversion of a substrate to product in a fluid environment—neither memorization of facts nor use of a computer program is going to help you. Thinking is always the hard part of most understanding.

The Concentration Factor(s)

If some kinetic event (let's stick with a substrate binding to an enzyme to form an enzyme–substrate complex) is truly a random (statistical) probability event that occurs only once in every (let's say) one million such events, it is easy to grasp that by increasing the numbers of substrate molecules per enzyme molecule (by increasing the concentration of substrate) we should experience that random (statistical) probability kinetic event in increasing numbers.

It is appropriate to remind you at this point that we are not talking about the entire enzyme mechanism where substrate is converted to product involving multiple kinetic events, but rather only about a single kinetic event where a substrate molecule binds to free enzyme to form an enzyme–substrate complex, as illustrated in Equation 13.1:

$$E + S \xrightarrow{k_1} ES \left[ES \underset{k_{-1}}{\longleftarrow} \xrightarrow{k_2} E + P \right]$$

$$(13.1)$$

(Where the rest of the kinetic mechanism associated with the conversion of substrate to product is shown apart from the kinetic event of interest.)

Based on Equation 13.1, at infinitely high concentrations of substrate (S), it is possible to convert all molecules of free enzyme (E) to the enzyme–substrate complex (ES); the event shown will exhibit first-order kinetics, and the concentration of the enzyme–substrate complex (ES) will increase in a linear fashion as the concentration of substrate (S) is increased. It should also be obvious that as one increases the concentration of substrate, the concentration of free enzyme (E) will decrease in a linear fashion, also exhibiting first-order kinetics.

There is an obvious assumption here that one is able to discriminate between free enzyme (E) and the enzyme bound to substrate in the form of ES. Given the ability to discriminate between the two forms of enzyme, AND in the absence of the rest of the events illustrated in Equation 13.1 as "the rest of the kinetic mechanism associated with the conversion of substrate to product," it should be relatively easy to calculate the value of the kinetic (rate) constant k_1 in Equation 13.1 by determining the slope of the line in Figure 13.1. The point of inflexion at substrate concentration 5 in Figure 13.1 is intended to illustration a concentration of substrate (S) at which all free enzyme is bound up as the enzyme–substrate complex. It should be obvious that the graph in Figure 13.1 is hypothetical; it is highly unlikely that you would ever obtain such a simple graph from real data. For example, the probability that you would ever choose a substrate concentration that would constitute the exact point of "saturation" of free enzyme is highly unlikely, and thus you should expect that the point of inflexion would be more "curvilinear" looking like a velocity versus substrate concentration data plot you have seen earlier. However, the data points after

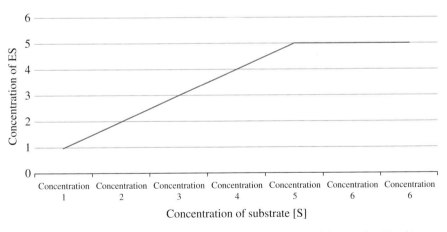

Figure 13.1 Hypothetical graph illustrating the rate of appearance of the "product" (in this case the enzyme–substrate complex as illustrated in Equation 13.1) by plotting the concentration of the enzyme–substrate complex ([ES]) as a function of the concentration of substrate ([S]). Theoretically the line would pass through zero, but this expectation is not shown here simply because the graph is hypothetical, and whether or not the line would pass through zero is speculative at best, but you can think more about that later. (*See insert for color representation of the figure.*)

that "curvilinear" point of inflexion would not represent some asymptotically approaching V_{max}, but rather an inability to form more enzyme–substrate complex because, in this case, there is no more free enzyme to convert to enzyme–substrate complex.

We also have a second concentration factor that we should consider from Equation 13.1, the concentration of free enzyme (E). What kinds of data plots might we expect for Equation 13.1 where the concentrations of free enzyme (E) vary when we hold the concentrations of substrate (S) at values that are always in "excess" (i.e., always saturating concentrations irrespective of the concentration of free enzyme)? The answer to this question is obvious: at saturating concentrations of substrate (S), adding more and more free enzyme will simply result in more and more enzyme–substrate complex (ES) being formed. Plotting "concentration of ES" (i.e., [ES]) versus concentration of enzyme E (i.e., [E]) will yield a linear (and "positive" data plot). It must thus be apparent to you by now that there is some other reason for dwelling on this question (and possible data plot(s)). The reason will be obvious if you look at Figure 13.2, in which I have provided a graphic with multiple data lines.

As you can see in Figure 13.2, there are three data plots showing how the concentrations of the enzyme–substrate complex [ES] (see Equation 13.1) change as the concentration of free enzyme [E] increases. One data line is linear out to a concentration of "4" for the free enzyme, but the other two lines begin to deviate from this linearity at a concentration of "1" for the free enzyme. One of the nonlinear data lines curves upward as the concentration of free enzyme increases, whereas the other nonlinear data line curves downward as the concentration of free enzyme increases. What conditions could lead to a nonlinear data plot of concentration of the enzyme–substrate complex versus

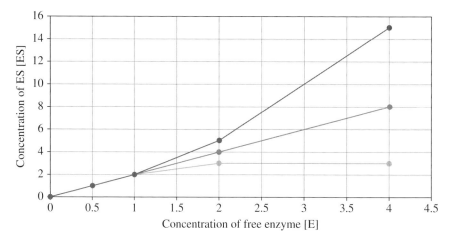

Figure 13.2 A hypothetical graph illustrating the change in the concentration of the enzyme–substrate complex [ES] versus change in the concentration of free enzyme [E] associated with Equation 13.1. (*See insert for color representation of the figure.*)

the concentration of free enzyme from Equation 13.1? Think about your answer(s), I'll wait. (Remember: we are still working under the assumption that the concentration of substrate [S] is at saturating levels at all concentrations of free enzyme.)

It really should not have taken you too long to suggest that the enzyme molecules began to aggregate (bind to each other?) as their concentrations increased, and in the "aggregated state" the kinetic constant k_1 changed for one or both (or more depending on the degree of aggregation) enzyme molecules. What do you think? Is that a good answer? It was the favorite of my students. I don't think it is a good answer. Why would I think it not to be a good answer? Want to think about it again? I'll wait.

Did you come up with the answer that the value of the rate constant k_1 doesn't matter? The association of substrate with free enzyme to form the enzyme–substrate complex may occur slower or faster (i.e., k_1 can decrease or increase for some reason), but the driving force is a concentration-dependent event and not a kinetic constant-dependent event. So, back to the original question. What conditions could lead to a nonlinear data plot of concentration of the enzyme–substrate complex versus the concentration of free enzyme from Equation 13.1 as illustrated in Figure 13.2?

Let's go back to Equation 10.7 to see if our thinking can be jogged. I'm repeating Equation 10.7 here as Equation 13.2 so you don't have to thumb back through this book. But be careful, not all of Equation 13.2 is relevant to answering the question posed—meaning we will edit Equation 13.2 to create Equation 13.3.

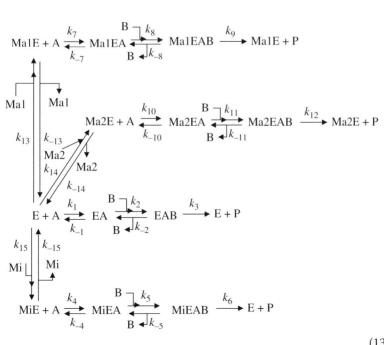

$$(13.2)$$

Actually, I will modify Equation 13.2 more than once, but let's start simple and work into complicated as we seek to answer the question posed. (Again, remember that we are not dealing with the conversion of substrate to product, but rather only the binding of substrate to free enzyme to form the enzyme–substrate complex—Equation 13.1—part of the overall kinds of enzyme mechanisms we have been dealing in in previous chapters.)

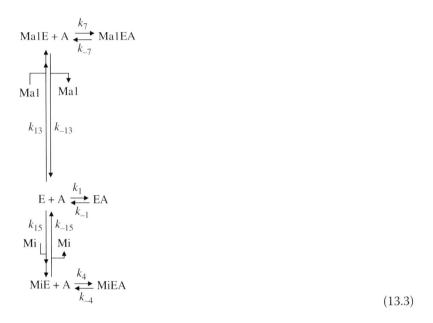

$$(13.3)$$

You will notice that I modified Equation 13.2 a great deal to get Equation 13.3. I deleted all parts of Equation 13.2 that went beyond formation of an enzyme substrate complex and I deleted the presence of a third "modifier" (the Ma2 modifier) so I could get you back to two possible "nonlinear data plots" as illustrated in Figure 13.3 and one linear data plot (the linear data plot we expected to get based on simple concentration-dependent forces associated with the formation of the enzyme–substrate complex in Equation 13.1).

You will also notice that unlike Equation 13.1, I left the formation of the enzyme–substrate complex a "reversible" event in Equation 13.3.

Now, I'm going to modify Equation 13.3 as Equation 13.4 to bring it into a form better suited to answering the question, "What conditions could lead to a 'nonlinear' data plot of concentration of the enzyme–substrate complex versus the concentration of free enzyme from Equation 13.1 as illustrated in Figure 13.3?".

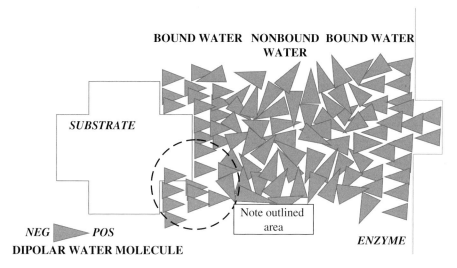

BOUND WATER   NONBOUND   BOUND WATER
WATER

SUBSTRATE

Note outlined
area

NEG ▶ *POS*
DIPOLAR WATER MOLECULE

ENZYME

Figure 13.3 Illustration of the role of bound water versus nonbound water in the translational movement of a molecule of substrate into the substrate-binding site of an enzyme [Figure 1.1 repeated]. (*See insert for color representation of the figure.*)

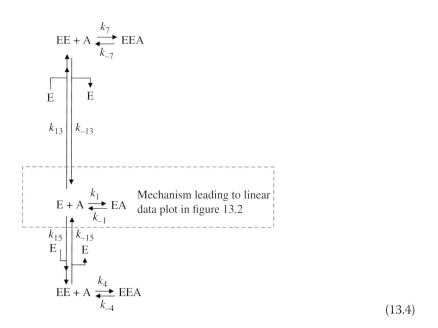

$$EE + A \underset{k_{-7}}{\overset{k_7}{\rightleftharpoons}} EEA$$

E   E

k_{13} k_{-13}

$$E + A \underset{k_{-1}}{\overset{k_1}{\rightleftharpoons}} EA \qquad \text{Mechanism leading to linear data plot in figure 13.2}$$

k_{15} k_{-15}
E   E

$$EE + A \underset{k_{-4}}{\overset{k_4}{\rightleftharpoons}} EEA$$

(13.4)

(Note that Equation 13.1 involved the substrate designated by S, not A as shown in Equation 13.4. I will shift back to designating substrate as S below for continuity, but A and S mean the same for our purposes in this discussion.)

Bending the Data Curve Downward in Figure 13.2

For purposes of illustrating how you might obtain the nonlinear data plot in Figure 13.2 where the data curve bends down as the concentration of free enzyme [E] increases could easily be explained by Equation 13.4. We don't really need "two" options for free enzyme (E) to bind to free enzyme (E) to form an EE complex, but I left in the "two" options so Equation 13.4 would look more similar to Equation 13.3 for your benefit. Also, remember that I've left the binding of E + A to form EA as reversible—something not present in Equation 13.1 at the beginning of this chapter. This "reversible" option will give you a different data plot in Figure 13.1 above, but we have finished with the discussion of why we get the data plot in Figure 13.1 so I left the binding reversible just so we can move forward with other issues.

If free enzyme can aggregate to a multiple enzyme complex (EE in this instance), but this multiple enzyme complex cannot combine with substrate (A) to form the enzyme–substrate complex (EA), it must be obvious to you that as the concentration of free enzyme increases, in Figure 13.2, less enzyme–substrate complex (EA) will be formed than if free enzyme (E) does not aggregate to form some multiple enzyme complex, and thus the obtained data plot will curve downward as the concentration of free enzyme increases.

It must also be obvious to you that how much the obtained data plot curves downward (if at all) will be (can be) driven by the valuations you might choose to assign to the kinetic constants k_1, k_{15}, and k_4 if you choose to make the enzyme aggregation event irreversible, and to kinetic constants, k_1, k_{-1}, k_{15}, k_{-15}, k_4, and k_{-4} should you choose to leave (make) the enzyme aggregation event reversible (as shown in Equation 13.4. For example, if you make k_1 much much greater (>>) than k_{15} and make the mechanisms irreversible, you can drive the point at which the linear and nonlinear downward-curving data plots in Figure 13.2 diverge to higher and higher concentrations of free enzyme. If you make k_1 and k_{15} approximately equal and keep the mechanism irreversible, you will find the two lines will begin to diverge from each other at even very low concentrations of free enzyme (E). The concentration of the enzyme–substrate complex (EA) can never approximate the concentration of free enzyme (E) because there will always be some free enzyme tied up either as the EE complex or as the EEA complex, and neither are the same as the EA complex. You really don't even need the EE complex to bind to substrate A to get a downward bending nonlinear data plot, but continuity is always better than making some quantum jump when dealing with proposed mechanisms in enzymology. Thus, the events illustrated in Figure 13.1 constitute more than a concentration-dependent event. They constitute some combination of concentration- and kinetic constant-dependent events. (Remember, we are operating under conditions, stipulated earlier, where substrate A is always at "saturating" concentrations with respect to the enzyme—meaning all free enzyme will be converted to the enzyme–substrate complex unless there are other places for the free enzyme to go.)

I'm sure that you appreciate that we could spend a whole chapter/book just looking at all possible options for how you can change the values of all possible combinations of kinetic constants in Equation 13.4 to make the nonlinear data plot appear in as many configurations as you can make up. So, although there is a lot more we could cover on this small topic, I think we should move on.

Bending the Data Curve Upward in Figure 13.2

The real question is: "Have I boxed myself into a corner trying to use Equation 13.4 to explain an upward bending nonlinear data plot in Figure 13.2?" Let's think about this matter. Can you see any way that increasing the concentration of free enzyme (E), based on the data plot in Figure 13.2 and on Equation 13.4, will lead to an increase in the concentration (amount) of the enzyme–substrate complex (EA)? Can combinations of kinetic (rate) constants be selected that will result in this outcome? Can you add to Equation 13.4 letting the EEA complex break down to free enzyme and the EA complex? Can you argue that forms of the enzyme complex such as EEA are basically the same as EA and thus the concentration of the enzyme–substrate complex ("EA") (or EA plus EEA) will increase in a linear fashion according to the data plot in Figure 13.2? This approach returns you to a concentration-driven event and it does not matter that there are kinetic constants routing enzyme in more than one pathway to the enzyme–substrate complex. All routes end up in the same place (just with different molecular configurations of the enzyme–substrate complex). One problem with this argument is that you get a linear data plot rather than an upward bending data plot.

It's OK to go up a wrong path sometimes, especially if you use the opportunity to sharpen your thinking skills. You probably understand that you can't drive free enzyme to some other form (any other form) that does not lead to the enzyme–substrate complex (EA) and gain increasing amounts of the enzyme–substrate complex with increasing concentrations of free enzyme using Equation 13.4 (or some modification of Equation 13.4).

How are you going to design a mechanism for the formation of the enzyme–substrate complex EA from free enzyme E and substrate A that will provide for a data plot that bends upward as shown in Figure 13.2? Let's assume that somehow you have to start with some form of enzyme (E*) that will yield more free enzyme E as you increase the concentration of enzyme E*. (Remember the basic premise of this book: That you are conducting experimentation and you get some upward bending data plot for the formation of some enzyme–substrate complex as you increase the concentration of enzyme and you need to figure out how you might have gotten such data.)

You should always start your thinking process by getting your ideas down in "black and white," so let's start there for now. Let's write out what will be Equation 13.5:

$$E^* \xrightarrow{k_0} E, \text{ and } E + A \xrightarrow{k_1} EA \tag{13.5}$$

or as rewritten in the format used in this book:

$$E^* \xrightarrow{k_0} E + A \xrightarrow{k_1} EA \qquad\qquad (13.5 \text{ part ii})$$

The most obvious implication of Equation 13.5 is that the enzyme E^* represents some form of "aggregated" enzyme (where I am using the term "free enzyme" to mean a single protein molecule) that will "deaggregate" into molecules of free enzyme (E) as you increase the concentration of enzyme E^* in the solution containing substrate A, and that this "deaggregation" will occur at a rate determined by the value of k_0. Equation 13.5 makes this event irreversible as it makes the formation of the enzyme–substrate complex irreversible.

So, what is wrong with Equation 13.5 relative to Figure 13.2? It doesn't allow for a linear data plot or for a curvilinear data plot that "turns down" as the concentration of free enzyme increases. Were you required to come up with an enzyme mechanism centered around Equation 13.1 that would allow not only a linear data plot, but two additional nonlinear data plots as illustrated in Figure 13.2? Remember the question: "What conditions could lead to a 'nonlinear' data plot of concentration of the enzyme–substrate complex versus the concentration of free enzyme from Equation 13.1?"

I think you have to agree that Equation 13.5 satisfies that question even though it lacks the ability to provide a linear data plot of concentration of the enzyme–substrate complex versus the concentration of free enzyme from Equation 13.1 in that Equation 13.1 is present in Equation 13.5. The former is just modified a bit to yield the latter.

I'm sure you get the point of this discussion: it is about learning to think about the questions you ask as well as how to arrive at the answers. Since there is rarely a single correct answer in enzymology, choose your questions carefully. The questions are often more important than the answers.

Other Factors Influencing Kinetic (Rate) Constants in Enzyme Mechanisms

Have you begun to wonder why a multi-enzyme complex such as S^* shown above in Equation 13.5 might begin to dissociate as the concentration of S^* (i.e., $[S^*]$) is increased? The normal process in biochemistry/enzymology (and, indeed, for many processes) is for molecules to aggregate as one increases their concentration.

We didn't influence kinetic (rate) constants in the enzyme mechanism in the examples above, we simply added new parts to some mechanism. However, the examples described give us the opportunity to think about what factors might be relevant in influencing kinetic (rate) constants associated with a given event in a single step in some enzyme mechanism that ultimately leads to the conversion of substrate to product. We seem to have asked another question: "What factor(s) might influence a multi-enzyme complex such as S^* to begin to dissociate as the concentration of S^* increases in concentration?" As protein molecules, enzymes influence the environment around them. Side groups on individual amino acids making up the enzyme molecule carry charges, either positive or negative, that vary depending on the pH and solvation characteristics of the solution in which the enzyme molecule is dissolved. The nature of these charges greatly influences not only the folding structure of the enzyme molecule, but also other molecules with which the enzyme comes into contact, and this influence extends not only to individual ions in solution, but also to other charged groups on other protein (enzyme) molecules with which the enzyme may associate. In addition, amino acid residues carrying more hydrophobic side groups influence protein structure by interacting with hydrophobic side groups of other amino acid residues in that protein and with hydrophobic side groups of other amino acid residues in other proteins. This interacting is one reason why the polarity of the solvent in which the enzyme is solubilized influences the structural organization of the enzyme, and how that enzyme interacts with other enzymes (and/or "substrates" and "modifiers," as discussed in previous chapters).

Imagine, for example, how fundamentally differently a given enzyme (protein) might be expected to behave when dissolved in plain water versus a modest solution of salt water versus a less modest solution of salty water. First of all, what constitutes "salt"? Salts come in a wide variety of forms and charge. Nearly all come with counter ions such that the positively and negatively charged counter ions balance each other; for example, sodium chloride (NaCl) has one positively charged sodium ion (Na^+) and one negatively charged chloride ion (Cl^-), whereas calcium chloride has one positively charged calcium ion (Ca^{2+}) and two negatively charged chloride ions (Cl^-). There are, of course, also molecular mass differences between sodium and calcium that represent differences when it comes to their interactions with other molecules, but charge is, in this particular case, more relevant and easier to talk about. For now let's stay with the plain water, modest salt water, and less modest salt water and their influence on enzymes/proteins (and ultimately kinetic/rate constants).

As positively charged species (cations), sodium ions will interact with negatively charged side groups of amino acids in some polypeptide chains (protein/enzyme) such as carboxyl groups of aspartic and glutamic acids (assuming, of course, that the side carboxyl groups of these amino acid residues are ionized), and these interactions will be reversible with their own kinetic (rate) constants. Like other reversible reactions we have considered, the greater the concentration of the sodium ion, the more likely the sodium ion is to associate with the negatively charged carboxyl group. If this ionized carboxyl group normally associates with some positively charged side group of amino acid residues such as lysine, arginine, or histidine in helping to establish a folding pattern of the polypeptide, replacing the interaction (association) with a sodium ion would be expected to alter the folding pattern of the polypeptide. Alternatively, if this ionizable carboxyl group resides within the substrate-binding site and is responsible for binding the substrate molecule (or for stabilizing/destabilizing a bond within that substrate), it becomes obvious that the presence or absence of a sodium ion in the enzyme environment will affect the role that enzyme plays in the binding of substrate or the conversion of substrate to product.

The question is: How does the presence of a sodium ion in the enzyme environment affect a given kinetic/rate constant associated with that enzyme? Does that sodium ion constitute a modifier such as we considered in previous chapters and equations (enzyme mechanisms)? If that sodium ion is interacting with a side group of an amino acid residue located in the binding/catalytic site of the enzyme and that interaction is reversible and substrate binding can compete with the sodium ion for interaction with the side group, does that make the sodium ion a "competitive modifier" (refer back to previous chapters for a competitive modifier definition)? We've always been taught that a competitive modifier (inhibitor if that makes it easier for you) was a "molecular analog" of the substrate in that it has always been "understood" that for a competitive modifier to interfere with the binding of a substrate molecule by an enzyme that modifier would need to be similar in structure to the substrate in order to bind in the binding/catalytic site of that enzyme. So, why have we never taught that a lowly sodium ion might act in a competitive manner similar to a molecular analog acting in a competitive manner with respect to binding of a substrate molecule in the substrate-binding site of an enzyme?

There are no rules here to guide you and I certainly don't want to start making rules as a way to answer questions. However, I think I can help you answer why we don't normally think of molecules like sodium ions as "competitive modifiers (inhibitors)" as described in the paragraph above. Firstly, ions in an aqueous solution of an enzyme and substrate are generally thought of as being like "water" molecules—they are ubiquitous in the aqueous solution and generally ignored in enzymatic reactions except where their roles are broad in nature affecting "everything equally." As an example, think of the effects of increasing concentrations of sodium chloride on an enzymatic reaction. The actions of the sodium and chloride ions affect virtually all aspects of the enzyme and substrate and thus we can ignore those actions with respect to specific enzyme mechanisms.

However (I love these "howevers"), this "difference" in treatment of sodium as a "modifier" versus some molecular analog of the substrate as a "modifier" (forgive me for the excessive use of quotation marks, but I want to emphasize that I am getting into near fiction territory) lets me probe your thinking process about factors that affect kinetic (rate) constants.

Much of this book has used modifiers to change enzyme mechanisms, but rather than changing kinetic (rate) constants on a given enzyme mechanism, the modifiers have relied on adding new enzyme mechanisms (with varying roles in changing the conversion of substrate to product) with their own (and normally distinct) kinetic (rate) constants. The modifiers don't change kinetic constants. Let me suggest to you that "modifiers," like sodium (but not limited to sodium), modify enzyme mechanisms by actually changing kinetic (rate) constants in the enzyme mechanism. Remember that earlier we talked about how kinetic (rate) constants were not really constant, but rather represented some average value of a range of values making up a given kinetic constant in a specific step in some overall enzyme mechanism. Factors such as ionic strength, polarity of the hydrating solvent (usually water), kind(s) of ions involved in producing ionic strength, and water-miscible solvents that alter the polarity of the hydrating solvent all participate in shifting the average value of a kinetic constant over the range of possible values and thus change the average value of a kinetic constant in an enzyme-mediated reaction. When you consider that there are multiple kinetic constants (and steps) in any given enzyme mechanism you might generate in consideration of the experimental data you collect, it is easy to accept that changes in the enzyme-mediated conversion of substrate to product over some range of pH values probably occur because you are changing one or more kinetic constants in your mechanism rather than adding new enzyme mechanisms as the solution pH changes. Remember that changes in pH of a solution represent changes in the concentration of the hydronium ion (H^+), and a hydronium ion is a positively charged ion (cation) capable of interacting with negatively charged side groups of amino acids making up the polypeptide/protein—the enzyme. The mechanisms (sequences of events) making up each step in your enzyme-mediated reactions do not change as a result of the "modifier"; rather the sum total of all the sequence of events making up the steps in your enzyme-mediated reactions changes.

Thus, Chapter 13 brings us back (finally) to Chapter 1 and our initial discussion of pseudo-thermodynamics and "self-assembly" events that are involved in, and are an integral part of, the enzymatic conversion of a substrate to a product. Solutes, ions, substrates, and enzymes serve to disrupt (disorder) the orderly structure of water and in so doing aid in the substrate approaching the enzyme (or vice versa). Let's go back to Figure 1.1 in Chapter 1, which I have copied here as Figure 13.3. However, let's then add things to Figure 13.3 to render a new Figure 13.4 to try to illustrate how other solutes, ions, and so forth might influence the conversion of substrate to product by altering the environment in which these latter two molecules are dissolved.

The point to be made by Figure 13.4 is that a solute or ion added to the water solution not only disrupts the ordered nature of the boundary water molecules in close proximity to substrate (in Figure 13.4), but if the solute or ion is charged it can also change the orientation of the dipolar water molecule with respect to other dipolar water molecules. The greater the concentration of such solutes/ions, the greater the degree of "disruption" with the result that it may take less "energy" to enable the substrate to approach the enzyme-binding site for that substrate. Remember, if the water in which substrate and enzyme are dissolved has an orderly (paracrystalline) structure, it possesses less entropy than water in which solutes/ions are dissolved. It thus takes less energy to disrupt the areas of boundary water when solutes/ions are present. There is, of course, also the possibility that in the substrate approaching an enzyme-binding site, the approaching tends to disrupt the orderly structure of boundary water, and with increasing entropy of the water molecules there is energy available to drive some chemical reaction (Why is nothing ever easy?). Remember, the early on attempt to cast the evaporation of water from an organic-rich solution as a driving force in the polymerization of nucleic acids/proteins/sugars? Remember, a negative Gibbs free energy $(-\Delta G)$ can be obtained if entropy (S) in Equation 1.1 is sufficiently large to make some chemical reaction occur spontaneously:

$$\Delta G = \Delta H - T\Delta S \tag{13.6}$$

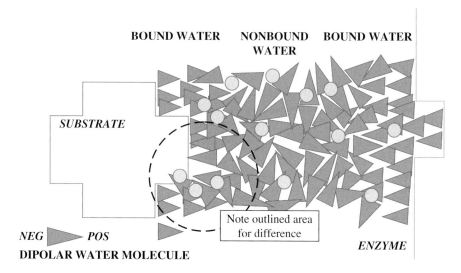

Figure 13.4 Illustration of the role of bound water versus nonbound water in the translational movement of a molecule of substrate into the substrate-binding site of an enzyme where other solute(s) is present to aid in the disruption of the structure of bound water in close proximity to both enzyme and substrate. (*See insert for color representation of the figure.*)

 I have thus taken a position that solutes such as ions (including changes in pH, ionic strength, etc.) differ in the way they modify enzyme reactions as compared to the modifiers we so vigorously discussed in earlier chapters. It is not important that this position be correct. Only that it causes you to think about the different ways in which other things present in the environment of your substrate and enzyme can impact on how you attempt to interpret your data. Avoid trying to make your data fit into enzyme mechanisms (reactions) that you read about or are taught in future lectures. Rather, try to take the route overly emphasized in this book, and come up with enzyme mechanisms (reactions—note the plurality) that fit your data.

> "And what does it live on?"
> "Weak tea with cream in it."
> "Supposing it couldn't find any?" she suggested.
> "Then it would die, of course."
> "But that must happen very often," Alice remarked thoughtfully.
> "It always happens," said the Gnat.
> Lewis Carroll, *Through the Looking Glass*

Index

Numbers in *italics* refer to figures.
Numbers in **bold** refer to tables.